PHILLIPSASTREID CORALS
FROM THE FRASNIAN (UPPER DEVONIAN) OF WESTERN CANADA:
Taxonomy and Biostratigraphic Significance

Inquiries: Monograph Publishing Program, NRC Research Press, National Research Council of Canada, Ottawa, Ontario K1A 0R6, Canada. Web site: www.monographs.nrc-cnrc.gc.ca

Photograph credits: Front cover outcrop: Late Frasnian mudmounds, Winnifred Pass, Alberta, photograph by author. Insets from top: *Macgeea parva* Webster, photograph by R. Strom; *Macgeea proteus* Smith, photograph by B. Rutley; *Macgeea pustulosa* n.sp., photograph by R. Strom. Back cover: Late Frasnian patch reef, Trout River, District of Mackenzie, photograph by author. Inset: Mid Frasnian Alexandra reef complex, Alexandra Falls, Hay River, District of Mackenzie, photograph by author.

Correct citation for this publication: McLean, R.A. 2005. *Phillipsastreid Corals from the Frasnian (Upper Devonian) of Western Canada: Taxonomy and Biostratigraphic Significance*. NRC Research Press, Ottawa, Ont. 109 pp.

A PUBLICATION OF THE
NATIONAL RESEARCH COUNCIL OF CANADA
MONOGRAPH PUBLISHING PROGRAM

PHILLIPSASTREID CORALS FROM THE FRASNIAN (UPPER DEVONIAN) OF WESTERN CANADA:
Taxonomy and Biostratigraphic Significance

Ross A. McLean

Anadarko Canada Corporation
Box 2595 Stn. M
Calgary, AB T2P 4V4, Canada

NRC·CNRC
NRC Research Press
Ottawa 2005

Printed in Canada on acid-free paper.

ISBN 0-660-19364-7
NRC No. 46847

Library and Archives Canada Cataloguing in Publication

McLean, Ross A.

Phillipsastreid Corals from the Frasnian (Upper Devonian) of Western Canada:
Taxonomy and Biostratigraphic Significance

Includes an absract in French.
Includes bibliographical references.
Issued by the National Research Council of Canada.
Available also on the Internet.
ISBN 0-660-19364-7

1. Rugosa – Canada, Western – Classification.
2. Corals, Fossil – Canada, Western.
3. Paleontology – Devonian.
4. Paleontology – Canada, Western.
I. National Research Council Canada.
II. Title.

QE665.M34 2005 563.6 C2004-980315-8

Contents

Abstract

Rugose corals of the Family Phillipsastreidae are abundant, diverse, and geographically widespread in the Frasnian (lower Upper Devonian) of western Canada. Species of the solitary genus *Macgeea* described here comprise *M. parva* Webster, 1889, *M. proteus* Smith, 1945, *M. telopea* Crickmay, 1962, *M. sorauﬁ* n. sp., and *M. pustulosa* n. sp. *Thamnophyllum* and *Peneckiella* are branching forms, with *Thamnophyllum* represented by the species *T. colemanense* (Warren, 1928), *T. tructense* (McLaren, 1959), *T. pedderi* n. sp., *T. cordense* n. sp. and *T. julli* n. sp., while *Peneckiella* includes *P. floydensis* (Belanski, 1928), *P. metalinae* Sorauf, 1972, *P. gracilis* n. sp., and *P. haultainensis* n. sp. Biostratigraphic distribution of these species is reviewed, together with that of previously described Canadian massive phillipsastreid species belonging to the genera *Phillipsastrea*, *Chuanbeiphyllum*, *Pachyphyllum*, *Smithicyathus*, and *Frechastraea.* The coral biostratigraphy is expressed in terms of the Montagne Noire conodont zonation and modified western Canada rugose coral faunal assemblages.

Résumé

Les coraux rugueux de la famille des Phillipsastreidae sont abondants, diversifiés et répandus sur le plan géographique dans le Frasnien (étage de base du Dévonien supérieur) de l'Ouest du Canada. Les espèces appartenant au genre solitaire *Macgeea* décrites ici sont *M. parva* Webster, 1889, *M. proteus* Smith, 1945, *M. telopea* Crickmay, 1962, *M. sorauﬁ* n. sp. et *M. pustulosa* n. sp. Les genres *Thamnophyllum* et *Peneckiella* sont des formes branchues, *Thamnophyllum* étant représenté par les espèces *T. colemanense* (Warren, 1928), *T. tructense* (McLaren, 1959), *T. pedderi* n. sp., *T. cordense* n. sp. et *T. julli* n. sp., tandis que *Peneckiella* inclut *P. floydensis* (Belanski, 1928), *P. metalinae* Sorauf, 1972, *P. gracilis* n. sp. et *P. haultainensis* n. sp. La répartition biostratigraphique de ces espèces est examinée, ainsi que celle d'espèces de phillipsastreidés canadiennes massives déjà décrites appartenant aux genres *Phillipsastrea*, *Chuanbeiphyllum*, *Pachyphyllum*, *Smithicyathus* et *Frechastraea.* La biostratigraphie des coraux est présentée suivant la zonation des conodontes de la Montagne Noire et les assemblages fauniques modifiés de coraux rugueux de l'Ouest du Canada.

Acknowledgments

The bulk of the material described here was collected by the writer over the years since 1979. Support of Amoco Canada Petroleum Company Limited (now BP Canada Energy) and Anadarko Canada Corporation for the outcrop and core work necessary to complete this study is gratefully acknowledged. Additional collections by the following geologists have also been utilized: A.S. Hedinger, R.K. Jull, R.G. McConnell, D.J. McLaren, E.W. Mountjoy, C.W. Stearn, P.S. Warren, E.J. Whittaker, and R.H. Workum.

I am grateful to A.E.H. Pedder of the Geological Survey of Canada for helpful discussions and the loan of certain literature and specimens for study. J.E. Sorauf of Binghamton University, New York, generously provided specimens from his collections of Iowa corals. Loans of type material in their care were arranged by T.E. Bolton (Geological Survey of Canada, Ottawa), B. Jones (University of Alberta, Edmonton), P.R. Hoover (Paleontological Research Institution, Ithaca, New York), B.F. Glenister and J. Golden (University of Iowa, Iowa City), and F.J. Collier (United States National Museum, Washington). T. Wrzołek (University of Silesia, Sosnowiec) kindly provided translations of some of the Polish literature.

Assistance of F. Grillo and G. Martin with preparation of thin sections is gratefully acknowledged. Text figures were drafted by D. Then and the photography is the work of R. Strom, B. Rutley, and K. Born. The manuscript benefited from careful review by A.E.H. Pedder and J.E. Sorauf.

Introduction

The Phillipsastreidae is a diverse and abundant family of rugose corals in the Frasnian of western Canada, comprising species with solitary, branching and massive growth forms. They occur in a variety of lithofacies, particularly in carbonate shelf margin and downslope facies, and may be significant frame builders in patch reefs on the shelf and "pinnacle" reefs on the slope. The 17 currently recognized species with massive coralla were described by McLean (1986, 1994*a*, *b*). In the present contribution, the five solitary and nine branching species are monographed to complete the current study of the family. With the completion of this taxonomic analysis and the establishment of a viable conodont and rugose coral biostratigaphic framework for the Frasnian of the Western Canada Sedimentary Basin (McLean and Klapper 1998), the biostratigraphic distribution of the family as a whole in the Frasnian of western Canada can now be reviewed.

Stratigraphic and Geographic Setting

The general sequence and correlations of coral-bearing horizons referred to in this paper are illustrated in Fig. 1. This diagram is based on earlier versions (McLean and Pedder 1984, Text-Fig. 2; McLean and Pedder 1987, Text-Fig. 2; McLean 1994*a*, Text-Fig. 2), but has been revised and expanded to reflect results of continuing study of the Frasnian succession of western Canada. The stratigraphy is correlated to the Montagne Noire conodont zonation of Klapper (1989) and a modified version of the rugose coral faunal sequence first described by McLean and Klapper (1998). Further study of the coral faunas has led to some slight changes in the original coral sequence of McLean and Klapper, those revisions being discussed in the Biostratigraphy section, below.

The areas discussed correspond to the stratigraphic columns of Fig. 1. The index map (Fig. 2) shows the location of small-scale maps in Figs. 3–14. The locality numbers on the maps, Figs. 3–14, are those of the Locality Register, where geographic, stratigraphic, and collection details are provided.

Root River area (Fig. 3)

The predominantly clastic sequence and fauna in the Frasnian of this area west of the Mackenzie River were reviewed by McLean and Pedder (1984, 1987), McLean (1994*a*) and McLean and Klapper (1998). The corals described herein were derived from an unnamed limestone unit within the Imperial Formation, largely correlatable with the Jean-Marie Member of the Redknife Formation, which outcrops and is present in subsurface to the south and east of this area. Species described here from the unnamed limestone include *Macgeea telopea* Crickmay and *Peneckiella metalinae* Sorauf.

Southern District of Mackenzie, Liard to Bouvier rivers, Kakisa Lake to Hay River (Figs. 4, 5)

The Frasnian sequence and faunas of this area have been discussed by McLean and Pedder (1984, 1987), McLean et al. (1987), McLean and Sorauf (1989), McLean (1994*a*) and McLean and Klapper (1998). The succession illustrated in Fig. 1 is a composite of outcrop and subsurface data. Outcrops in the Hay River – Kakisa Lake area are mainly of the lower part of the sequence, while the upper part is exposed mostly on tributaries flowing into the Mackenzie River.

On Hay River *Macgeea proteus* Smith and *M. soraufi* n. sp. are described from members B and C of the Hay River Formation, and *Peneckiella floydensis* (Belanski) makes its first appearance in the section in Member C. The reef complex represented by the Alexandra Member

Fig. 1. Stratigraphic terminology and generalized correlations, Frasnian of western Canada. Asterisks indicate horizons bearing coral specimens described in this paper. Rugose coral faunas and their correlation to conodont zones modified from McLean and Klapper (1998).

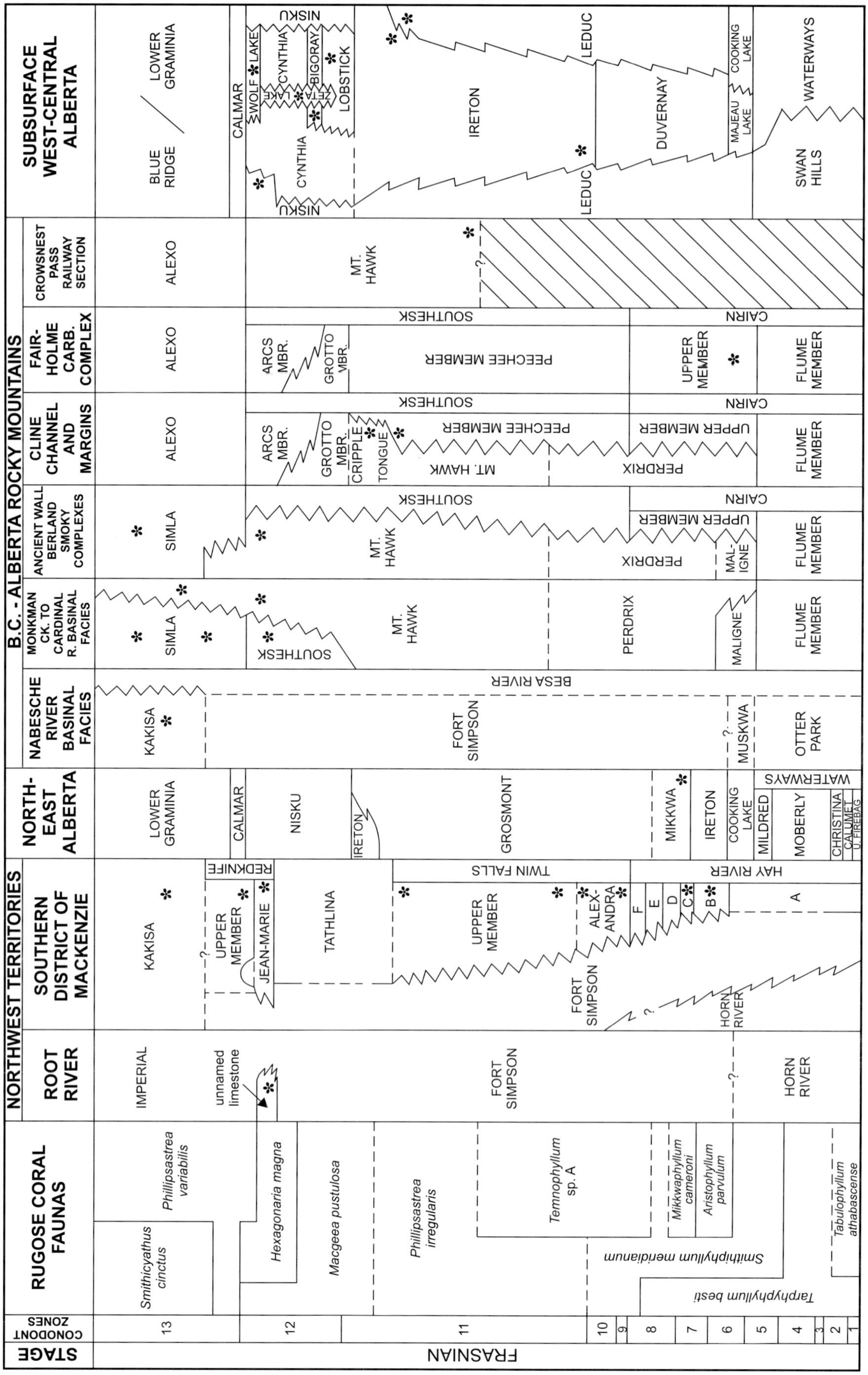

Fig. 2. Index map of part of western Canada, showing location of small-scale maps, Figs. 3–14. (For Map area A, see Fig. 3; Map area B, Fig. 4; Map area C, Fig. 5; Map area D, Fig. 6; Map area E, Fig. 7; Map area F, Fig. 8; Map area G, Fig. 9; Map area H, Fig. 10; Map area I, Fig. 11; Map area J, Fig. 12; Map area K, Fig. 13, Map area L, Fig. 14).

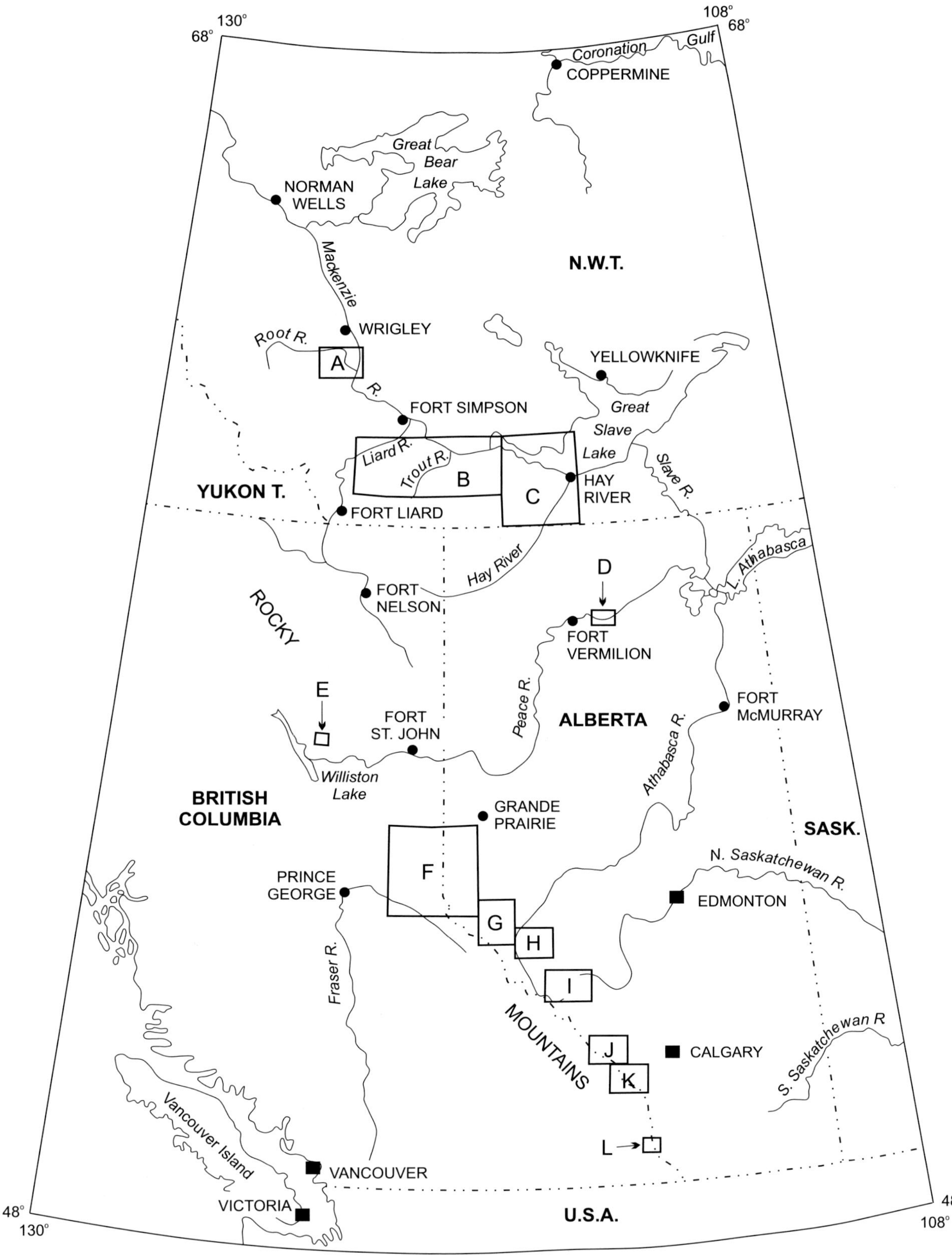

Fig. 3. Locality map, Root River area, District of Mackenzie, NWT (Map area A, Fig. 2).

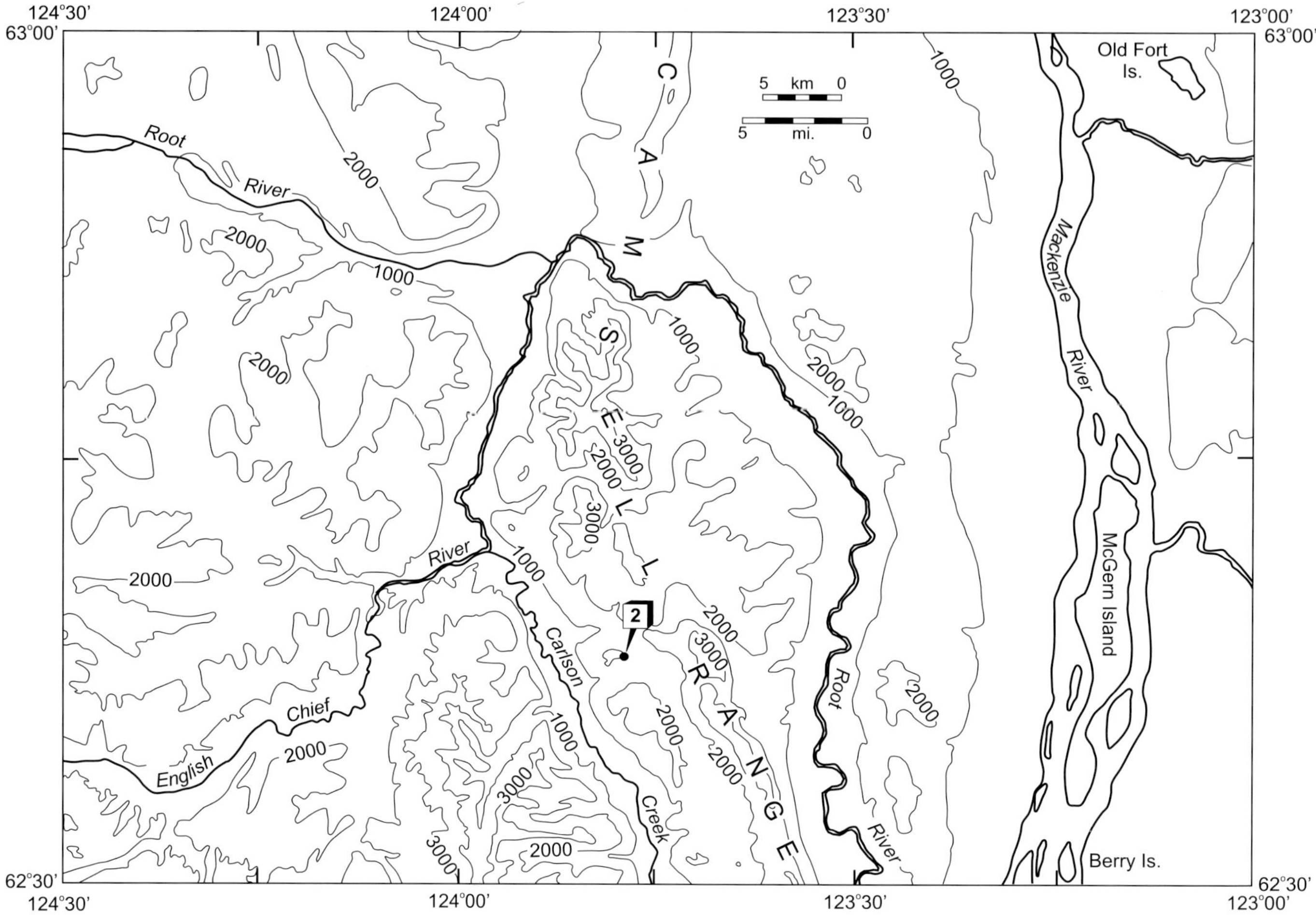

Fig. 4. Locality map, Blackstone to Bouvier rivers area, District of Mackenzie, NWT (Map area B, Fig. 2).

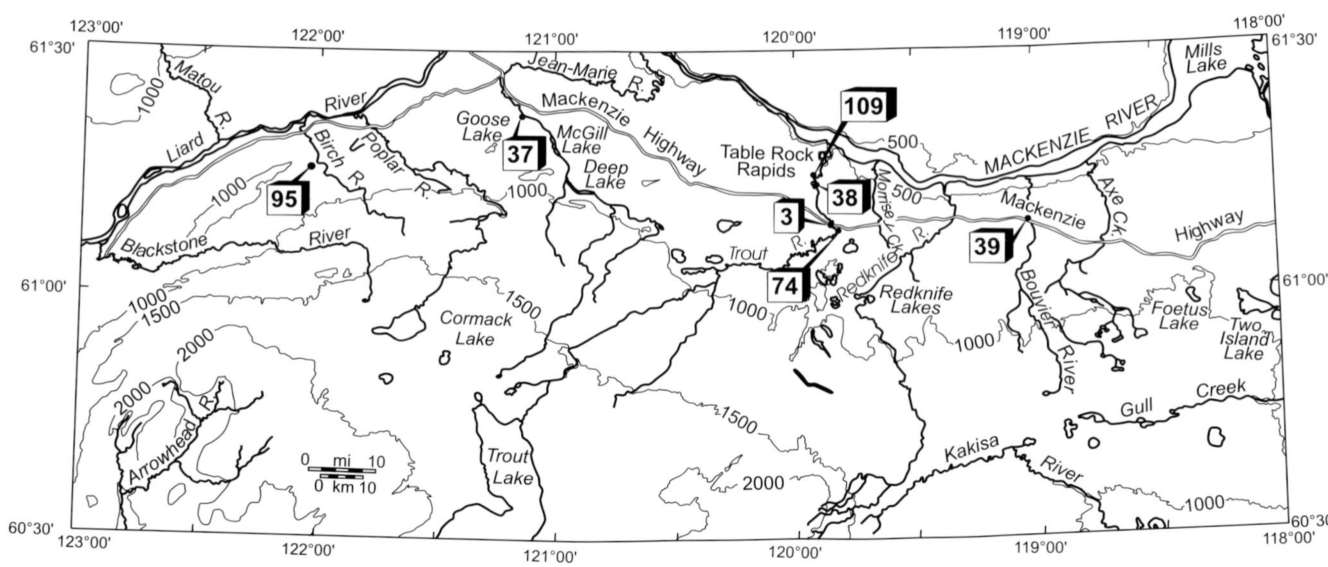

Fig. 5. Locality map, Kakisa Lake to Hay River area, District of Mackenzie, NWT (Map area C, Fig. 2).

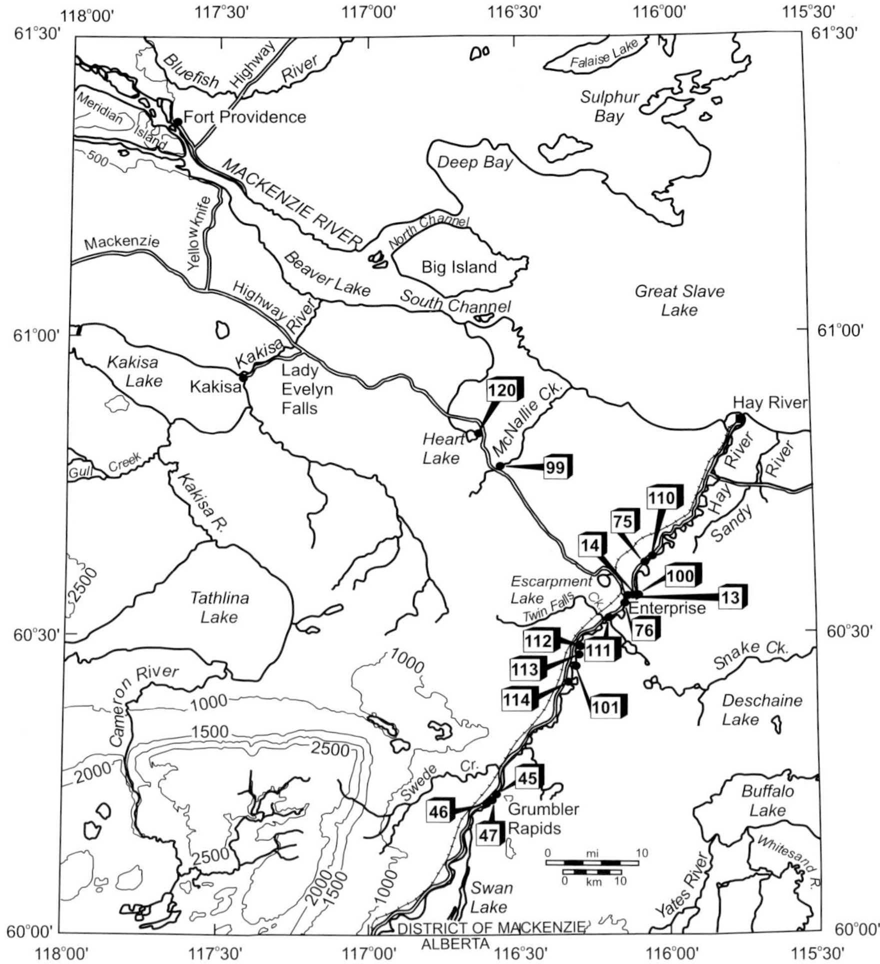

of the Twin Falls Formation contains *M. soraufi* in its basal, platform beds and *Thamnophyllum cordense* n. sp. is described from fore-reef debris facies near Heart Lake. Further material of *Peneckiella floydensis* is present in the succeeding upper member of the Twin Falls Formation on Hay River.

Rugose corals are also described here from the late Frasnian portion of the sequence in the western part of the area. The Jean-Marie Member of the Redknife Formation has *Macgeea pustulosa* n. sp., and *Peneckiella metalinae* Sorauf, while *Macgeea pustulosa* is also present in the overlying upper member of the Redknife. *Thamnophyllum tructense* (McLaren) is described from the youngest Frasnian unit in the area, the Kakisa Formation.

Northeast Alberta (Fig. 6)

The stratigraphic column depicted in Fig. 1 for this area is a composite of outcrop and subsurface data. The biostromal carbonates of the Mikkwa Formation outcropping on Peace River near Vermilion Falls were discussed by McLean and Pedder (1984), McLean (1994*a*) and McLean and Klapper (1998). An addition to the coral fauna cited in those papers is *Peneckiella floydensis* (Belanski).

Rocky Mountains, Nabesche River basinal facies (Fig. 7)

McLean (1994*a*) and McLean and Klapper (1998) have given reviews of the Frasnian stratigraphy and faunas of this area of northeast British Columbia. *Thamnophyllum tructense* (McLaren) is described here from beds assigned to the late Frasnian Kakisa Formation.

Rocky Mountains, Monkman Creek to Cardinal River basinal facies (Figs. 8–10)

Stratigraphy and faunas of the Frasnian off-reef succession of this region have been discussed by McLean and Pedder (1984, 1987), McLean (1994*a*) and McLean and Klapper (1998). The upper Mt. Hawk Formation, comprising argillaceous limestones and calcareous shales equivalent to the upper Southesk Formation of

Fig. 6. Locality map, Vermilion Falls area, northeast Alberta (Map area D, Fig. 2).

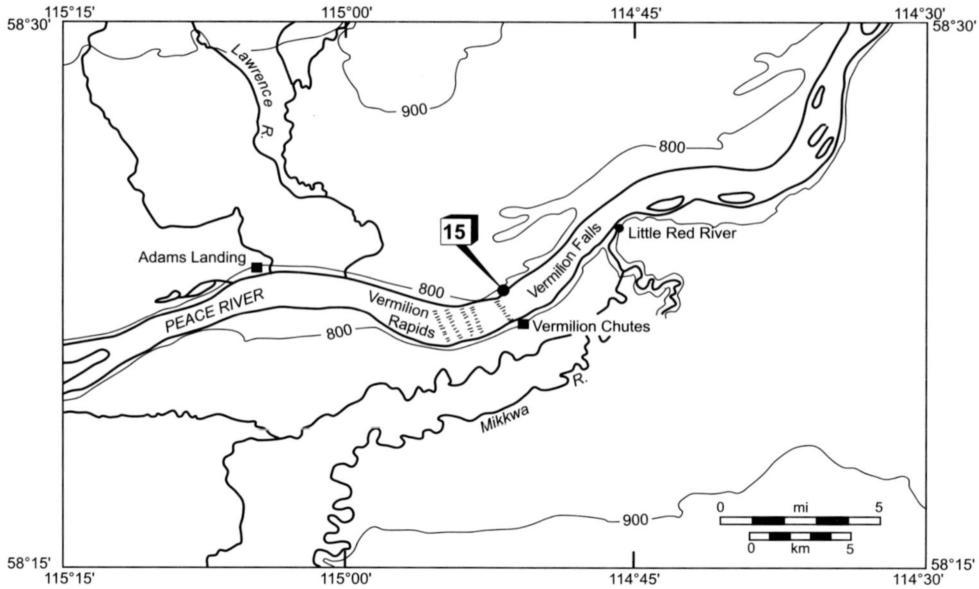

Fig.7. Locality map, Nabesche River area, northeast British Columbia (Map area E, Fig. 2).

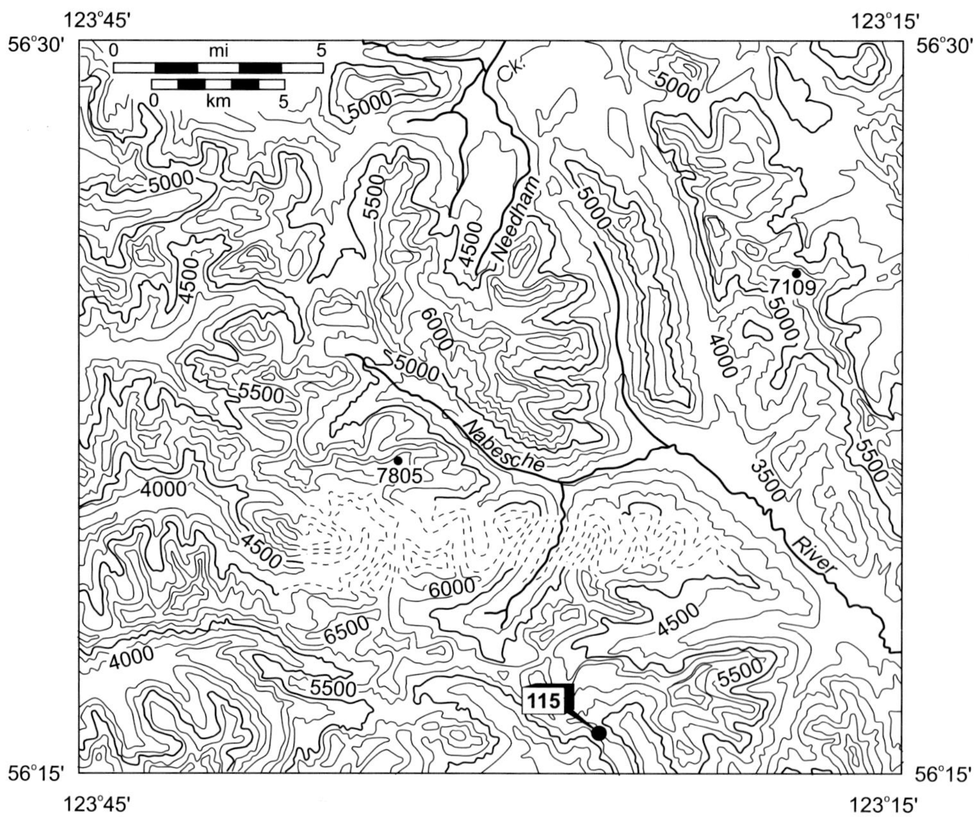

the reefal facies, commonly has a rich rugose coral fauna, with a preliminary list given by McLean (1994*a*, p. 47). Species described in the present work from this interval include *Macgeea pustulosa* n. sp., *Thamnophyllum colemanense* (Warren), and *T. pedderi* n. sp. From undifferentiated upper Southesk Formation carbonates equivalent to these beds, *Thamnophyllum julli* n. sp. is described. North of Cardinal River, the latest Frasnian basinal sequence is represented by richly fossiliferous biostromal carbonates. These beds have had a complex history of stratigraphic nomenclature, reviewed by McLean (1994*a*, pp. 47–49). Following

Fig. 8. Locality map, Monkman Creek to Winnifred Pass area, British Columbia-Alberta Rocky Mountains (Map area F, Fig. 2).

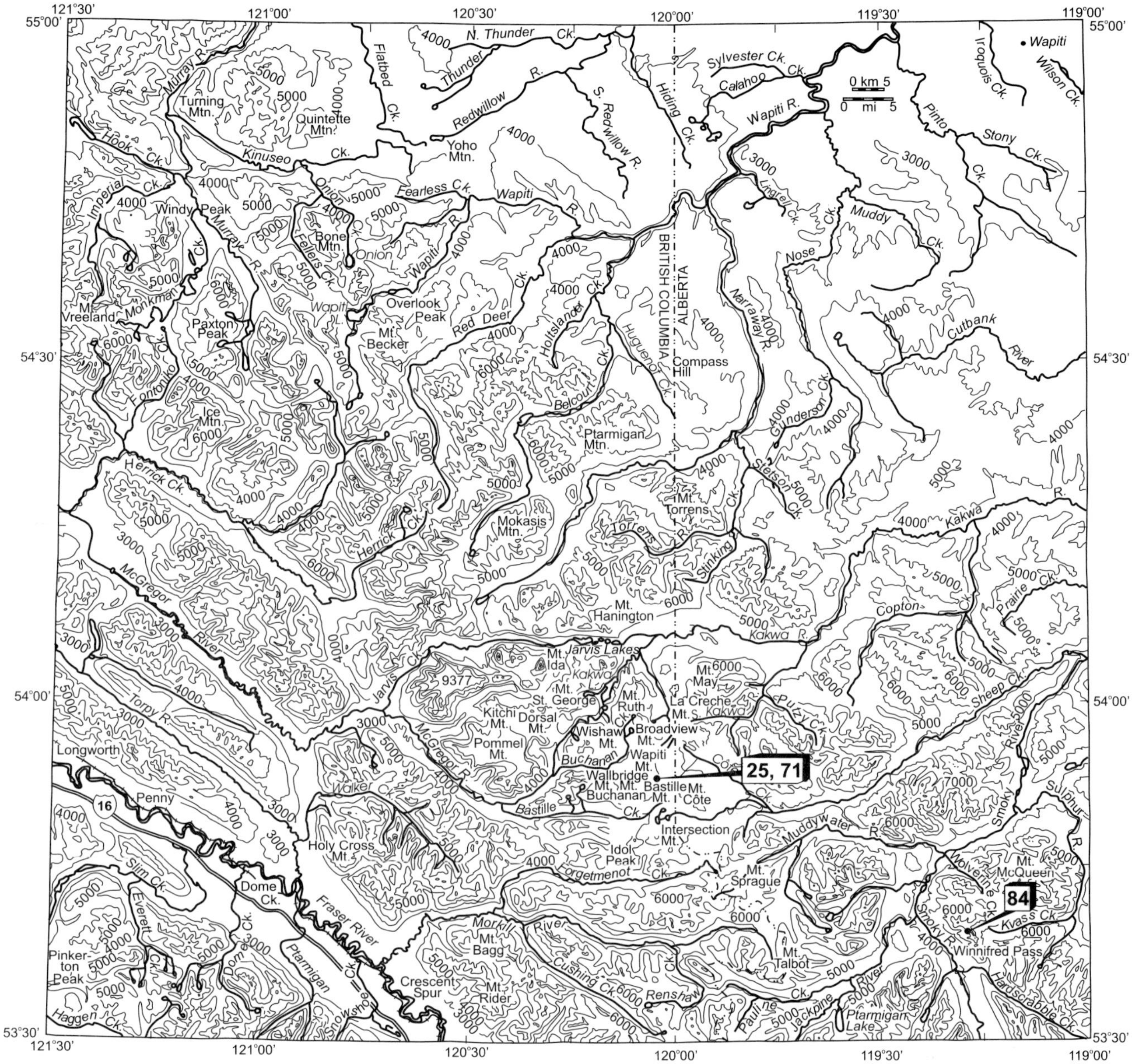

the conclusions of that study, the term Simla Formation is used here for those strata. Supplementing the faunal lists of McLean and Sorauf (1989, p. 385) and McLean (1994*a*, p. 49), the species *Thamnophyllum tructense* (McLaren), *T. colemanense* (Warren), and *Peneckiella haultainensis* n. sp. are described below. In some cases, the highest Frasnian beds in the basinal sequences are significantly argillaceous and cannot be easily differentiated from the underlying Mt. Hawk Formation. Such strata are regarded as an upward extension of the Mt. Hawk and the rich coral fauna at one such locality (Whitehorse Creek, Fig. 10, Loc. 78) was listed by McLean (1994*a*, p. 49). *Peneckiella haultainensis* n. sp. (*Peneckiella* sp. in that list) is described here from that locality.

Rocky Mountains, Ancient Wall, Berland River, and Smoky carbonate complexes and margins (Figs. 8, 9)

Stratigraphy and faunas occurring at the margin of the Ancient Wall carbonate complex northwest of Jasper, Alberta have been reviewed by McLean and Pedder (1984, 1987), McLean (1994*a*) and McLean and Klapper (1998). The southeast margin of the complex at Mt. Haultain contains a rich rugose coral fauna collected by

Fig. 9. Locality map, Ancient Wall area, Alberta-British Columbia Rocky Mountains (Map area G, Fig. 2).

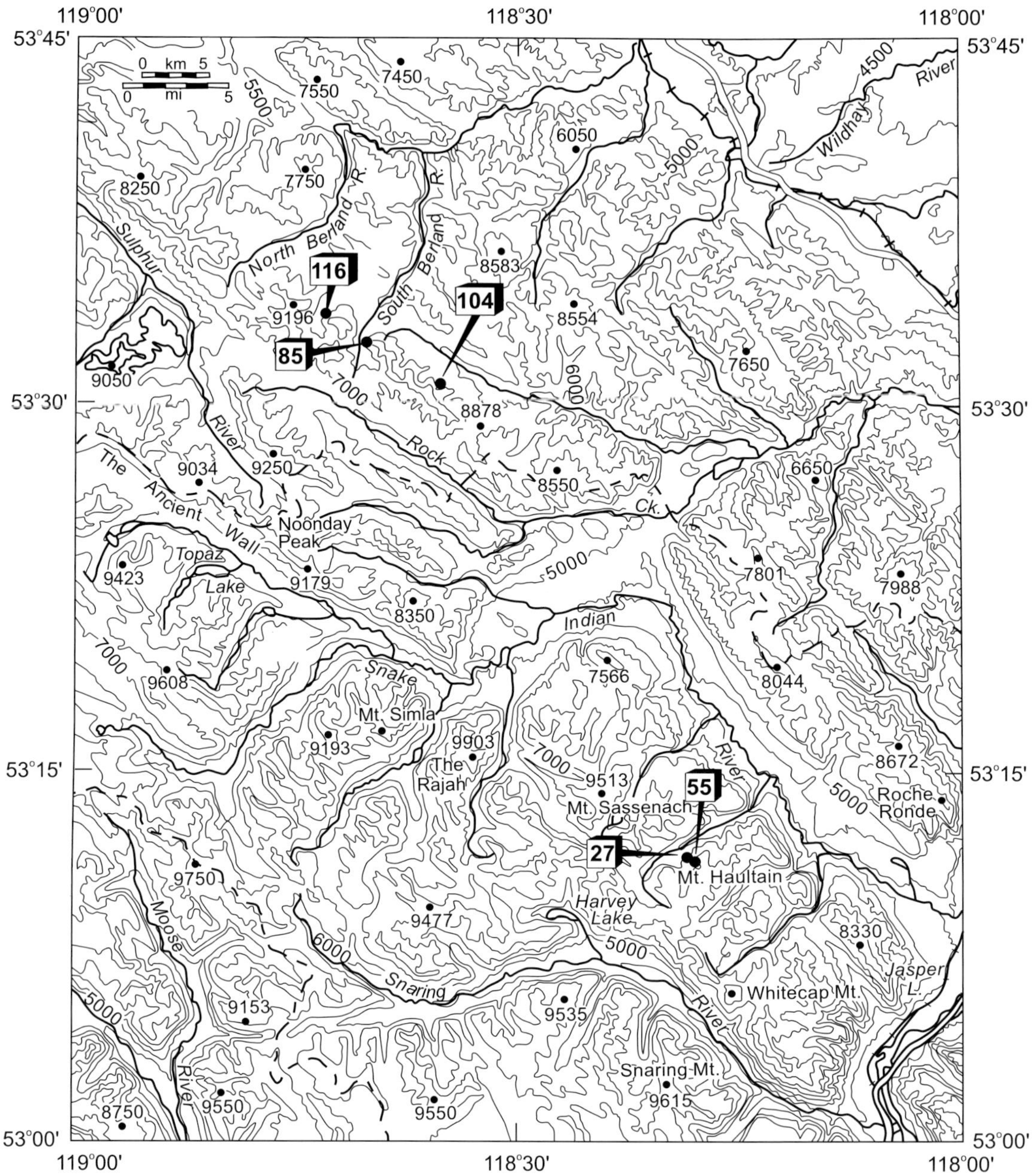

the late R.K. Jull. At this locality, the upper Southesk Formation, just below the Simla Formation, contains *Thamnophyllum julli* n. sp., while more distal, slightly older strata of the Southesk carry *T. colemanense* (Warren). *Macgeea telopea* Crickmay is described from the upper Mt. Hawk Formation, in beds overlying the upper Southesk Formation. The biostromal carbonates of the Simla Formation, which prograde basinward over the Southesk and Mt. Hawk formations, yielded *Peneckiella haultainensis* n. sp. at this locality.

A small buildup north of the Ancient Wall complex exposed in the Persimmon Range is referred to as the Berland carbonate complex (see Mountjoy 1978, Fig. 2). McLean (1994*a*) and McLean and Klapper (1998) reviewed its stratigraphy and faunas and elements of that assemblage described here include *Thamnophyllum colemanense* (Warren) from the upper Southesk Formation and *T. tructense* (McLaren) in the overlying Simla Formation. At the southeast margin of the complex, *Macgeea pustulosa* n. sp. is described from the upper Southesk beds immediately below onlapping strata of the basinal Mt. Hawk Formation.

To the northwest of the Ancient Wall and Berland complexes, a small buildup was referred to informally as the Smoky complex and briefly discussed by McLean and Klapper (1998, p. 533). It has yielded corals at its

Fig. 10. Locality map, Roche Miette to Mt. Mackenzie area, Alberta Rocky Mountains (Map area H, Fig. 2).

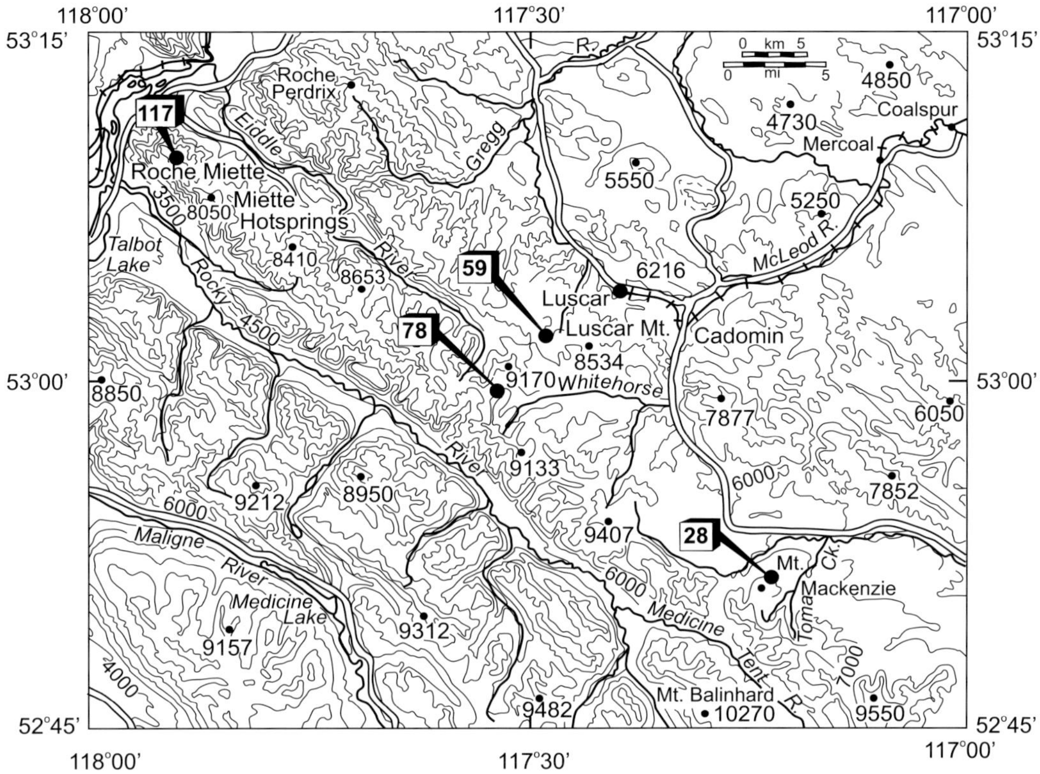

Fig. 11. Locality map, North Saskatchewan River area, Alberta Rocky Mountains (Map area I, Fig. 2).

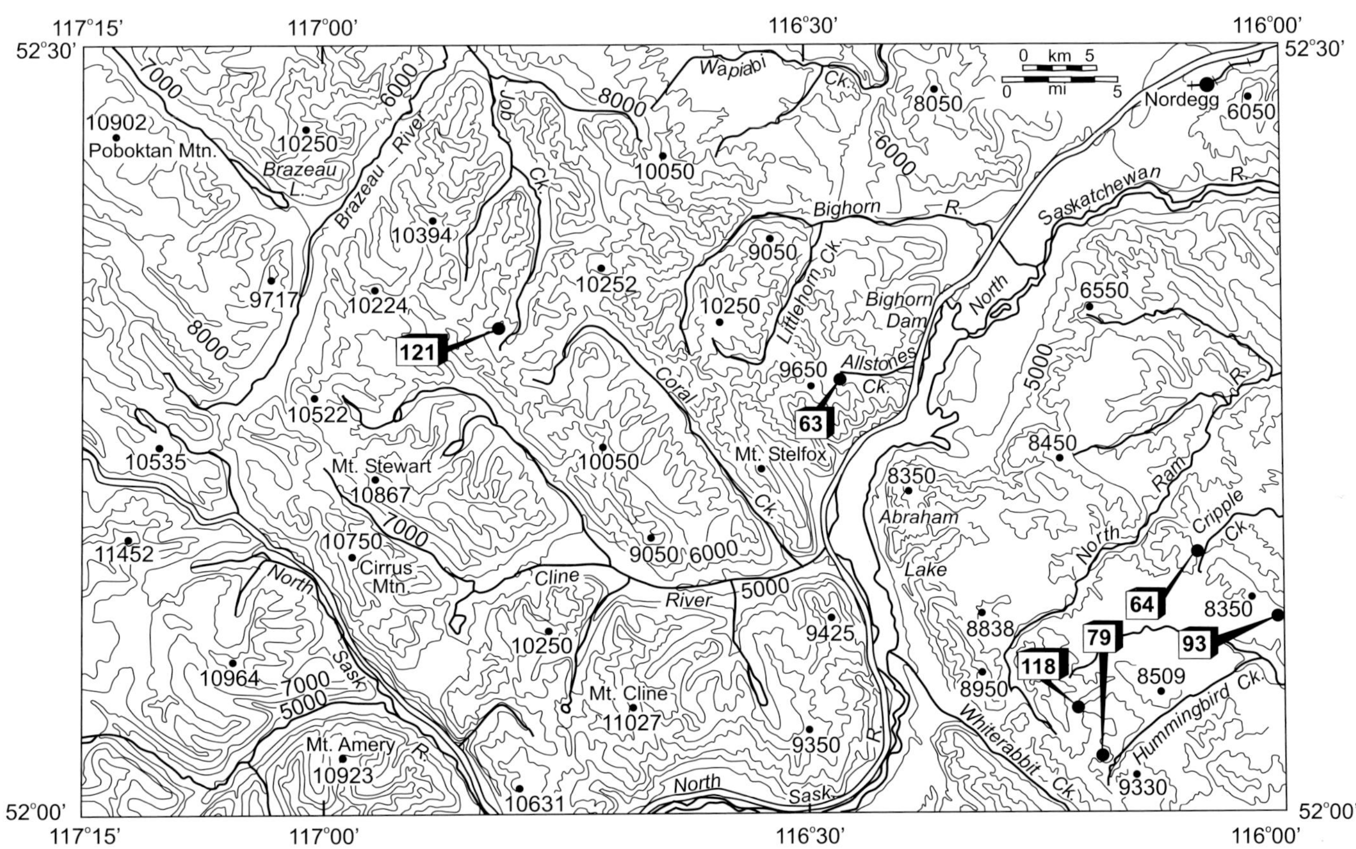

Fig. 12. Locality map, Banff area, Alberta-British Columbia Rocky Mountains (Map area J, Fig. 2).

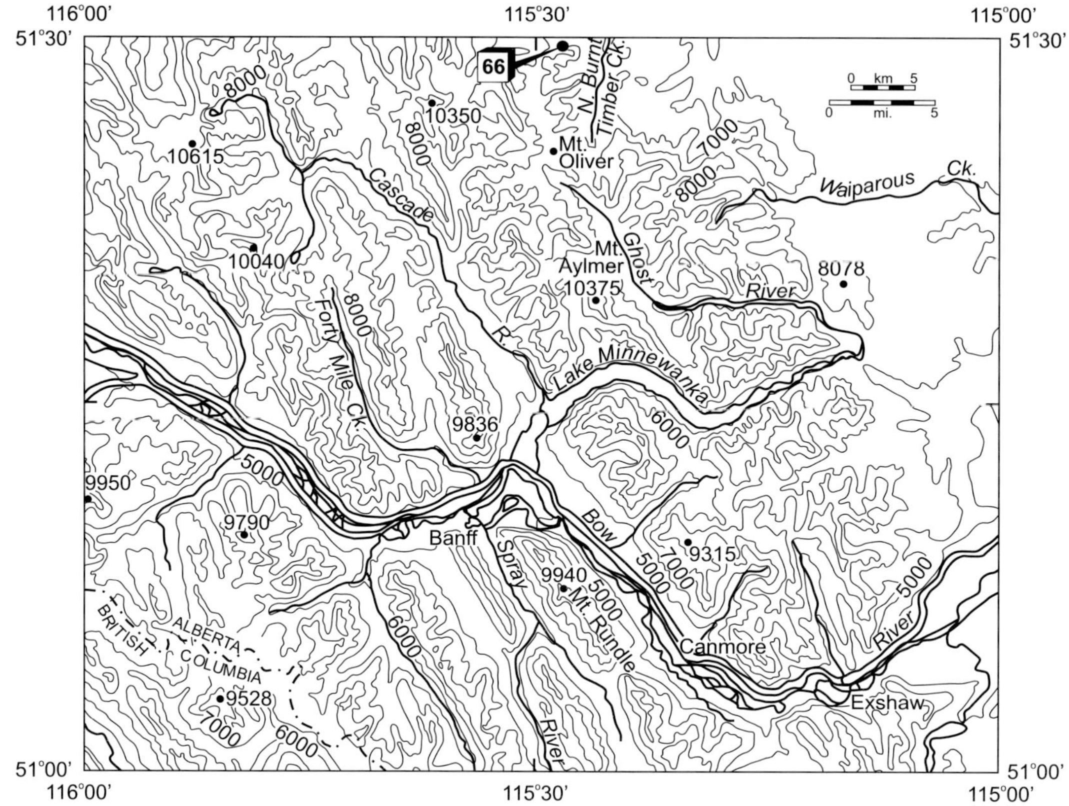

southeast margin near Childear Mountain (McLean 1994*a*, p. 50) and *Peneckiella metalinae* Sorauf is described here from strata of the upper Mt. Hawk or lower Simla formations.

Rocky Mountains, Cline Channel and margins (Fig. 11)

The stratigraphy and faunas of the Cline Channel, a basinal feature separating the Southesk Cairn and Fairholme carbonate complexes, and of the margins of those complexes, have been reviewed by McLean and Pedder (1987), McLean and Sorauf (1989), McLean (1994*a*) and McLean and Klapper (1998). *Peneckiella gracilis* n. sp. and *P. floydensis* (Belanski) are present in limy beds of the uppermost Peechee Member, Southesk Formation, while from the overlying, onlapping calcareous shales and argillaceous limestones of the Mt. Hawk Formation, locally referred to as the Cripple Tongue, *P. gracilis*, *P. floydensis*, and *Thamnophyllum pedderi* n. sp. are described.

Rocky Mountains, Fairholme carbonate complex (Figs. 12, 13)

McLean and Pedder (1984, 1987) and McLean and Klapper (1998) discussed the stratigraphy and faunas of this large carbonate complex outcropping southeast of the Cline Channel. In general, the strata of the complex are pervasively dolomitized, but locally some units are only mildly dolomitized or, rarely, preserved as limestone, and in those cases corals may be well enough preserved to permit taxonomic evaluation. *Thamnophyllum pedderi* n. sp. is described from the upper member of the Cairn Formation near North Burnt Timber Creek and from undifferentiated Cairn Formation at Mt. McDougall.

Rocky Mountains, Crowsnest Pass area, southwest Alberta (Fig. 14)

Frasnian stratigraphy of the reef and basinal sequence of the Flathead Range – Crowsnest Pass area of southwest Alberta has been reviewed by Workum and Hedinger (1992), building on the initial studies of Warren (1928), de Wit and McLaren (1950), and Price (1962, 1965). The only area relevant to the present work is in the Crowsnest Pass itself, site of Warren's original collection of *Thamnophyllum colemanense*, recollected by the present writer in 1993. The incomplete, faulted, basinal section along the railway line north of Crowsnest Lake (Fig. 14, Loc. 119), described by Warren (1928) and de Wit and McLaren (1950) and depicted in Fig. 1, contains approximately 110 m of argillaceous and silty dolomites and limestones of the Mt. Hawk

Fig. 13. Locality map, Kananaskis area, Alberta-British Columbia Rocky Mountains (Map area K, Fig. 2).

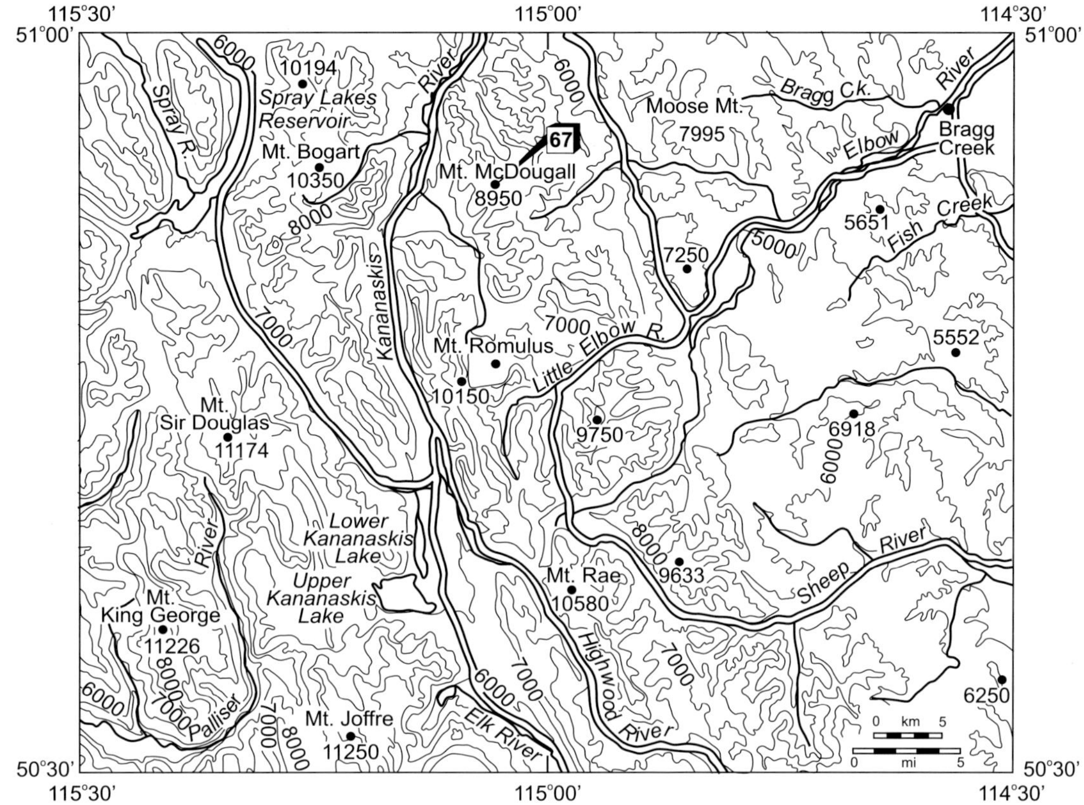

Fig. 14. Locality map, Crowsnest Pass area, Alberta-British Columbia Rocky Mountains (Map area L, Fig. 2).

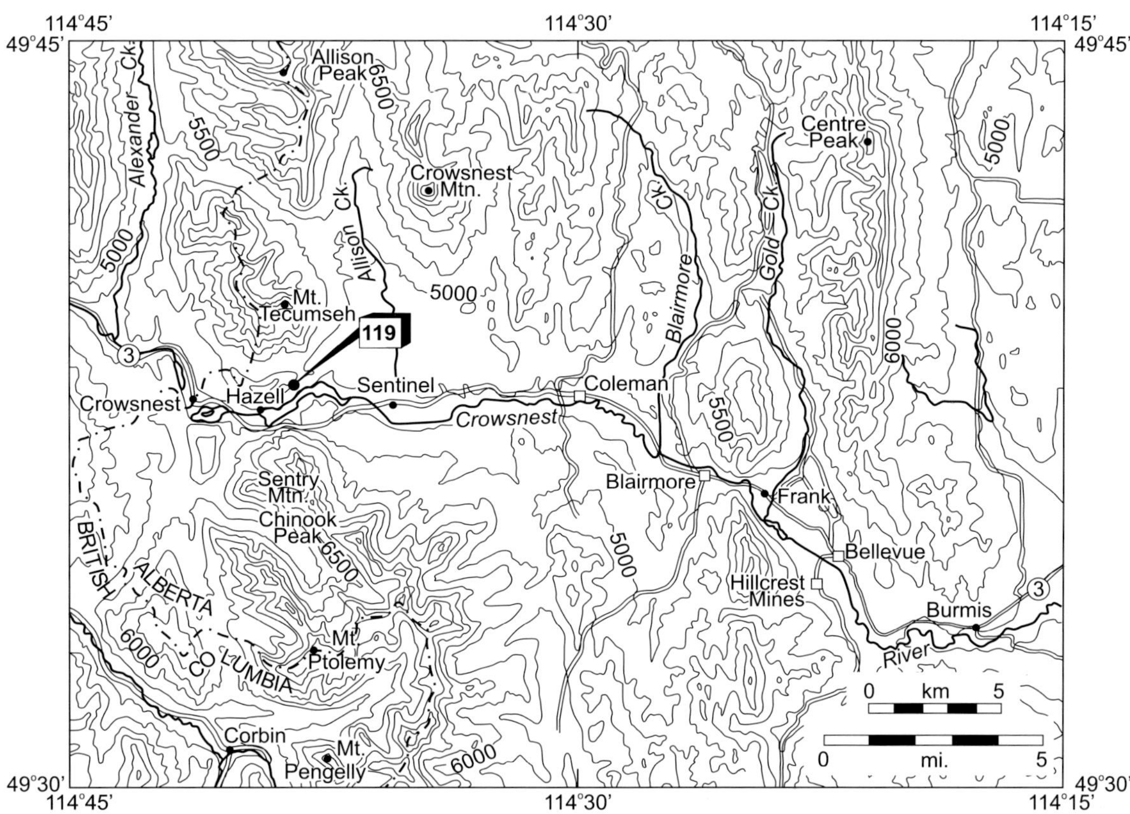

Fig. 15. Simplified subsurface paleogeographic map, mid-late Frasnian, west-central Alberta, showing numbered locations of cored wells cited in the text. Shaded areas indicate mid Frasnian Leduc carbonate complexes; dashed line is margin of late Frasnian Nisku carbonate shelf outlining Winterburn Basin. 1: 13-19-69-8W6M, 2: 9-27-70-23W5M, 3: 7-32-65-19W5M, 4: 6-3-67-15W5M, 5: 10-3-68-14W5M, 6: 4-29-57-3W5M, 7: 12-1-57-3W5M, 8: 11-20-58-22W4M, 9: 4-18-58-21W4M, 10: 5-5-58-24W5M, 11: 10-32-58-23W5M,

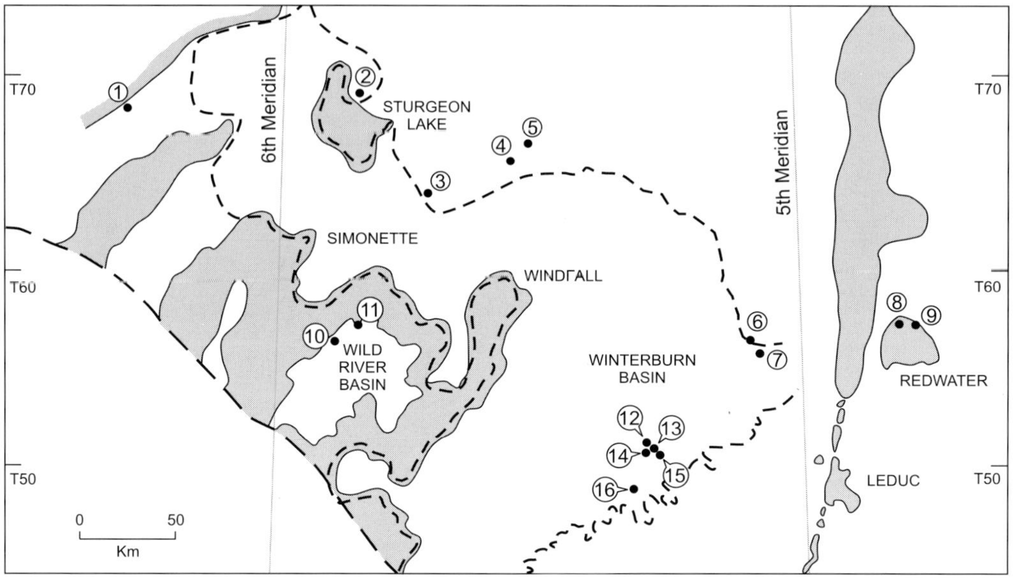

Formation, underlying the Alexo Formation. Near the base of the Mt. Hawk exposure, thin biostromes contain *T. colemanense* (possible topotypes) and abundant *Peneckiella metalinae* Sorauf, as noted by McLean and Klapper (1998, p. 542), and material of both species from this locality is described in the present work.

Subsurface west-central Alberta (Fig. 15)

The subsurface Frasnian succession in Alberta, the focus of intensive hydrocarbon exploration for over 60 years, has received extensive treatment in the published literature. Parts of the sequence relevant to the distribution of rugose corals were reviewed in some detail by McLean and Klapper (1998) and the reader is referred to that publication for history of development of stratigraphic nomenclature and interpretation. An assemblage of carbonate bank complexes, ranging from early Frasnian (Swan Hills Formation), mid Frasnian (Leduc Formation) to late Frasnian (Nisku Formation), is present in the region, separated, onlapped, and in some cases interbedded with different stages of basinal infill. A simplified paleogeographic reconstruction of the area at two stages of bank development (Leduc and Nisku formations) is depicted in Fig. 15. Within the areas of basin fill, smaller "pinnacle" reefs of both Leduc and Nisku age are also commonly developed. The subsurface stratigraphic column in Fig. 1 is a composite one and is based on the more extensive correlation chart of McLean and Klapper (1998, Fig. 1). The units of interest here are the Ireton, Leduc, and Nisku formations. The Ireton, a shaly, basinal succession, tends to have a considerable carbonate component near the reef complexes, and may host abundant rugose coral faunas in such settings. It onlaps and is interbedded with reefal carbonates of the Leduc Formation. The latter is commonly dolomitized, but there are some cases where it is still preserved primarily as limestone, and while the faunas are dominated by stromatoporoids, rugose corals may also be present. The overlying Nisku Formation is composed mainly of shallow water dolomites and evaporites in eastern and southern Alberta, but in west-central Alberta the Nisku bank passes to the largely shale and argillaceous limestone succession of the Winterburn Basin (Fig. 15). At the southern margin of that basin and in downslope "pinnacle" reefs, the Nisku has been subdivided into a series of members (Exploration Staff, Chevron Standard Ltd. 1979) and shown in the subsurface column of Fig. 1. On the north side of the basin, shelf carbonates are present only in the upper Nisku, which has not been subdivided.

Macgeea parva Webster is described here from the uppermost Leduc Formation of the Redwater complex and from the immediately overlying Ireton Formation (Well locs. 8, 9, Fig. 15). In the Wild River Basin, *Peneckiella floydensis* (Belanski) is described from the upper Leduc Formation of the Colt "pinnacle" reef (Well loc. 10, Fig. 15), and *Thamnophyllum pedderi* n. sp. is present in the upper Ireton Formation (Well loc. 11, Fig. 15).

The remainder of the subsurface specimens dealt with in the present work are from the Nisku Formation. *Macgeea pustulosa* n. sp. occurs in shaly equivalents of the lower Nisku bank (Well loc. 1, Fig. 15). The undifferentiated upper Nisku on the north side of the Winterburn Basin yielded *Thamnophyllum colemanense* (Warren) (Well locs. 2, 4, 7, Fig. 15) and *T. pedderi* n. sp. (Well locs. 3, 5, 6, Fig. 15). On the south side of the basin, the distal ramp facies of the Lobstick and Wolf Lake members has *T. colemanense* (Well locs. 12, 16, Fig. 15), while *T. pedderi* is described from the Bigoray Member (Well loc. 15, Fig. 15). From the "pinnacle" reef facies of the Zeta Lake Member, *Thamnophyllum julli* n. sp. (Well locs. 13, 14, Fig. 15) and *Peneckiella metalinae* Sorauf (Well loc. 14, Fig. 15) are described. It should be noted that these last two species, together with species of *Smithiphyllum* Birenheide, 1962 are the dominant frame builders of the Zeta Lake Member reefs.

Biostratigraphy

An initial attempt at a biostratigraphic zonation of Frasnian rugose corals in western Canada was given by Pedder *in* Moore (1993, Table 4D.1). This scheme was substantially modified by McLean and Klapper (1998), who introduced a sequence of 11 rugose coral faunal assemblages, linked to 13 conodont zones developed initially in the Montagne Noire region of southern France (Klapper 1989). Subsequent study of the coral faunas has led to some minor modifications to the coral sequence of McLean and Klapper (1998). *Macgeea pustulosa* n. sp. is introduced here for the species informally referred to as *Macgeea* sp. B by McLean and Klapper (1998), and is used here to designate that faunal assemblage. The underlying assemblage, approximately equivalent to conodont zone 11, was named the *Peneckiella densa* fauna by McLean and Klapper (1998), but *P. densa* (Smith) is regarded in the present work as a junior synonym of *P. floydensis* (Belanski). Material now regarded as representative of *P. floydensis* in western Canada has a considerably longer range than that originally understood for *P. densa* by McLean and Klapper (1998) (see Fig. 16, herein), and so an alternative species has been designated for the interval approximating to conodont zone 11. *Phillipsastrea irregularis* (Webster and Fenton, 1924) ranges from about the top of the range of *Smithiphyllum meridianum* McLean and Pedder, 1987 to the base of *Macgeea pustulosa* n. sp. (*Macgeea* sp. B of McLean and Klapper 1998, Fig. 1), is widespread in western Canada, and is an appropriate replacement for *P. densa* in designating the coral assemblage.

The conodont zonation and revised coral faunal sequence are shown in Figs. 1 and 16. Fig. 1 illustrates their relationship to the stratigraphic successions relevant to the current study, while Fig. 16 shows the range of western Canadian species of the Phillipsastreidae expressed in terms of the conodont and coral sequence. Species of the massive genera of the Phillipsastreidae are included in Fig. 16 and in the discussion below as their biostratigraphy was not analyzed (and coral assemblage sequence not developed) at the time of their study (McLean 1986, 1994*a*, *b*). *Macgeea* is represented by five species in the Frasnian of western Canada. *Macgeea proteus* Smith and *M. soraufi* n. sp. both appear first in member B of the Hay River Formation on Hay River, southern District of Mackenzie, with *M. proteus* ranging into the Alexandra Member and *M. soraufi* into the overlying lower upper member of the Twin Falls Formation in that area. Both species also occur in the lower Leduc Formation of the Alberta subsurface, while *M. proteus* is recognized in the Mikkwa Formation of northeast Alberta. *Macgeea parva* Webster appears to have a restricted range, being known only in the upper Leduc – basal Ireton formations of the central Alberta subsurface. *M. pustulosa* n. sp., name giver to the *M. pustulosa* fauna, is a widespread species, whose range is close to that of conodont zone 12. It occurs in outcrop in the Jean-Marie Member and lower upper member of the Redknife Formation, southern District of Mackenzie, and equivalent strata of the upper Mt. Hawk Formation of the Alberta Rocky Mountains. In subsurface west-central Alberta it is common throughout the Nisku Formation, primarily in distal bank margin facies of that unit. *Macgeea telopea* Crickmay is a rare species, known only from the unnamed limestone unit of the Imperial Formation in the Root River area, District of Mackenzie, and from the upper Mt. Hawk Formation at the margin of the Ancient Wall carbonate complex of the Alberta Rocky Mountains.

The genus *Thamnophyllum* also has five species recognized in the Frasnian of western Canada. *T. pedderi* n. sp. appears first and is a long ranging form, extending from the upper Cairn Formation of the Fairholme carbonate complex, as high as the upper Mt. Hawk Formation (zone 12 equivalent, *M. pustulosa*/*H. magna* coral faunas) in the Alberta Rockies. It is particularly common in the subsurface, occurring in the Ireton, upper

Fig. 16. Biostratigraphic distribution of species of the Phillipsastreidae in western Canada, expressed in terms of the rugose coral faunal sequence and conodont zonation modified from McLean and Klapper (1998).

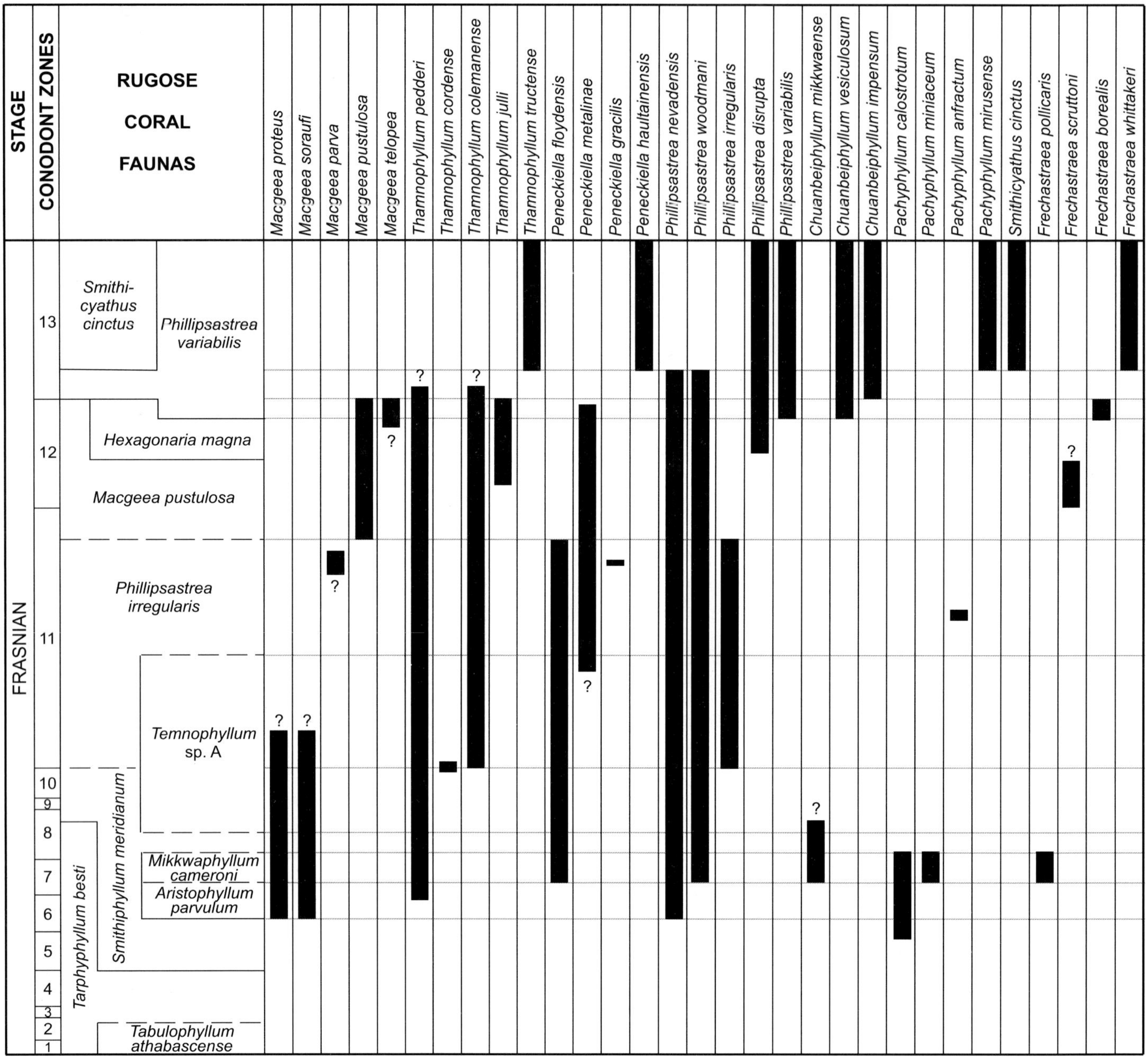

Leduc, and especially Nisku formations in Alberta, and the Jean-Marie Member of the Redknife Formation of northeast British Columbia. *T. cordense* n. sp. is distinctive, but very rare, known only from fore-reef debris facies of the Alexandra Member (approximate zone 10 equivalent) of southern District of Mackenzie. In outcrop, *T. colemanense* (Warren) is found in the uppermost Twin Falls Formation on Hay River, southern District of Mackenzie, ranging up to the upper Mt. Hawk and equivalent strata of the upper Southesk and lower Simla formations of the Alberta Rockies. Like *T. pedderi*, it is very common in the Alberta subsurface, occurring in the upper Duvernay, Ireton, upper Leduc, Nisku and Calmar formations, being particularly abundant in distal facies of the Nisku. *T. julli* n. sp. appears to be quite facies dependent, occurring in outcrops of the upper Southesk Formation near the margin of the Ancient Wall carbonate complex, and as an important frame builder in the "pinnacle" reefs of the Zeta Lake Member, Nisku Formation, in the Alberta subsurface. The youngest representative of *Thamnophyllum* in western Canada is *T. tructense* (McLaren). It is restricted to the *P. variabilis/S. cinctus* coral faunas (approximately zone 13) of the late Frasnian in outcrop of the Kakisa

Formation of southern District of Mackenzie and northeast British Columbia, and the Simla and equivalent uppermost Mt. Hawk formations of the Alberta Rockies.

Four species of *Peneckiella* are recognized here in the Frasnian of western Canada. *P. floydensis* (Belanski) appears earliest, in member C of the Hay River Formation of southern District of Mackenzie, equivalent strata of the Mikkwa Formation in northeast Alberta, and the upper Cairn Formation of the western margin of the Fairholme carbonate complex of the Alberta Rockies. In southern District of Mackenzie it ranges up to the top of the Twin Falls Formation, and in the Rockies to the Cripple Tongue of the Cline Channel area (zone 11). In subsurface it extends from equivalents of the Mikkwa Formation (lower Grosmont Formation of subsurface terminology) through the Leduc Formation. *P. metalinae* Sorauf is a very widespread species, in outcrop ranging from the uppermost Twin Falls Formation to the Jean-Marie Member and equivalent unnamed limestone unit in District of Mackenzie, while in the Alberta Rockies it is found in the Cripple Tongue and upper Mt. Hawk Formation (upper zone 11 – zone 12). In the Alberta subsurface it occurs in the upper Leduc, Ireton, upper Grosmont, and Nisku formations. *P. gracilis* n. sp., on the other hand, is a geographically restricted species, known only from the upper Peechee Member and Cripple Tongue of the Cline Channel area of the Alberta Rockies. The youngest western Canadian representative of *Peneckiella* is *P. haultainensis* n. sp., restricted to the Kakisa Formation of southern District of Mackenzie and the Simla and equivalent uppermost Mt. Hawk formations of the British Columbia and Alberta Rockies.

Considering now the massive phillipsastreids of the western Canadian Frasnian, described by McLean (1986, 1994*a*, *b*), the genus *Phillipsastrea* itself is the most diverse and widespread, with five species currently recognized. *P. nevadensis* Stumm is the first to appear, in subsurface occurring in the Cooking Lake Formation (zones 5–6) and extending through the Ireton, Grosmont, Leduc, and Nisku formations into the Blue Ridge Member of the Graminia Formation (lower zone 13). It is also present in the subsurface Jean-Marie Member of northeast British Columbia. In outcrop it is also very long ranging, known from members B and C, Hay River Formation, upper Twin Falls Formation, and Jean-Marie Member of southern District of Mackenzie. Elsewhere in District of Mackenzie it occurs in the unnamed limestone unit of the Root River area and Imperial Formation (zone 12) of the Norman Wells area, central Mackenzie valley. Strata of the Rockies of Alberta and British Columbia which contain this species include the upper Southesk and lower Simla formations of the carbonate bank facies and equivalent basinal beds of the upper Mt. Hawk Formation. *P. woodmani* (White) is also a widespread and long ranging species. In outcrop it is found in members C and D of the Hay River Formation, and the Alexandra and upper members of the Twin Falls Formation in southern District of Mackenzie. It is also known from the unnamed limestone unit in the Root River area. In the Alberta and British Columbia Rockies it is recognized in the upper Cairn Formation of the Fairholme carbonate complex, Cripple Tongue of the Cline Channel area, and upper Mt. Hawk to lower Simla formations of the basinal successions. Similarly, it is widespread in the subsurface, ranging from the Mikkwa Formation through the Grosmont, Ireton, and Nisku formations to the lower Blue Ridge Member. *P. irregularis* (Webster and Fenton) first appears in the Leduc Formation of the Alberta subsurface and is also in the upper Peechee Member, Southesk Formation at the margin of the Smoky complex and Cripple Tongue of the Alberta Rockies. In southern District of Mackenzie it occurs in beds of the upper Twin Falls Formation (approximately coeval with the Cripple Tongue) at Grumbler Rapids on Hay River. *P. disrupta* McLean is present in the unnamed limestone unit in the Root River area and equivalent Jean-Marie Member of southern District of Mackenzie. It is found also in the upper Southesk Formation and equivalent basinal strata of the upper Mt. Hawk Formation, together with the overlying Simla Formation of the British Columbia and Alberta Rocky Mountains. In subsurface it is rare, being recognized from only one locality in the Wolf Lake Member of the upper Nisku Formation. The youngest representative of *Phillipsastrea* in western Canada is *P. variabilis* (Sorauf), which occurs earliest in the upper member of the Redknife Formation of southern District of Mackenzie, and in approximately equivalent beds of the Imperial Formation in the central Mackenzie valley (uppermost zone 12). It is widespread in the Kakisa Formation of southern District of Mackenzie and northeast British Columbia, together with the Simla Formation and equivalent beds of the upper Mt. Hawk Formation in the British Columbia – Alberta Rockies (zone 13). In the Alberta subsurface it is also found in the Blue Ridge Member of the Graminia Formation.

Chuanbeiphyllum mikkwaense McLean is the earliest representative of that genus to appear in the western Canadian Frasnian. It occurs in both outcrop and subsurface sections of the Mikkwa Formation in northeast Alberta (*Mikkwaphyllum cameroni* coral fauna). The other two species of *Chuanbeiphyllum* have only been found in the late Frasnian. *C. vesiculosum* (Smith) makes its first appearance in high zone 12 equivalent strata of the Imperial Formation in the central Mackenzie valley, and is common in the Kakisa Formation of southern District of Mackenzie and northeast British Columbia. It is also widespread in the Simla Formation and equivalent upper Mt. Hawk strata of the British Columbia – Alberta Rockies. *C. impensum* McLean is a relatively rare form and seems to be restricted to zone 13 age strata of the Simla Formation in the British Columbia – Alberta Rockies and coeval Blue Ridge Member of the Alberta subsurface.

Pachyphyllum is an uncommon genus in western Canada, with four morphologically variable species currently recognized. The oldest form is *P. calostrotum* (Crickmay), known only from the Cooking Lake and Mikkwa formations of subsurface eastern Alberta. *P. mineaceum* McLean occurs in the Mikkwa Formation in outcrop of northeast Alberta, while *P. anfractum* McLean is only recorded from the upper Peechee Member, Southesk Formation of the Fairholme complex in the Alberta Rocky Mountains. Youngest representative of the genus is *P. mirusense* McLean, from the late Frasnian Simla Formation of the British Columbia Rockies.

Smithicyathus has only one species recognized in western Canada, *S. cinctus* (Smith). However, it is quite widely distributed in the latest Frasnian (name giver for the *S. cinctus* coral fauna, Figs. 1, 16), and is known from the Kakisa Formation, southern District of Mackenzie, Simla Formation of the British Columbia – Alberta Rockies, and in the Blue Ridge Member of the Alberta subsurface.

The final massive phillipsastreid genus to be considered here is *Frechastraea*, with four species in western Canada. *F. pollicaris* McLean appears first, with occurrences in member C, Hay River Formation (*M. cameroni* fauna) of southern District of Mackenzie. *F. scruttoni* McLean is a rare component of the Mt. Hawk Formation fauna (zone 12) in the Alberta Rockies, but is more common in the Lobstick Member of the Nisku Formation and shaly equivalents in the central Alberta subsurface. *F. borealis* McLean is also a rare form and is known only from high zone 12 age strata of the Imperial Formation in the central Mackenzie valley. Youngest species is *F. whittakeri* (Smith), with records in the Kakisa Formation, southern District of Mackenzie, and Nabesche River area of northeast British Columbia, together with the Simla Formation of the British Columbia Rockies.

Systematic Paleontology

Morphological terminology used is that of Hill in the two editions of the Treatise (1956, 1981), and the reader is referred to those volumes for clear descriptions and illustrations of skeletal characters relevant to the phillipsastreids. Repositories for specimens cited in the text are indicated by the following prefixes for specimen numbers:

GSC — Type Collections of the Geological Survey of Canada, Ottawa

UA — Paleontological Collections, Department of Geology, The University of Alberta, Edmonton

PRI — Paleontological Research Institution, Ithaca, New York

SUI — Department of Geology, The University of Iowa, Iowa City

Family PHILLIPSASTREIDAE Roemer, 1883
1883 Phillipsastraeidae Roemer, p. 389

Diagnosis. Corallum solitary, branching or massive. Septa in the dissepimentarium typically show fusiform dilation, with mainly full fans of trabeculae, generally rhipidacanthine. Horseshoe, peneckielloid, and globose dissepiments, in varying combinations, occur adjacent to the tabularium, while outer dissepiments may be globose, elongate, flat, or concave. Tabulae are complete and incomplete, flat, variably arched, or weakly concave (modified from McLean 1989, p. 239).

Discussion. Massive genera assigned to the Phillipsastreidae were reviewed by McLean (1989, 1994*a*, *b*). Solitary genera included here in the family are *Macgeea* Webster, 1889 and possibly *Protomacgeea* Różkowska, 1956, while the branching forms are *Thamnophyllum* Penecke, 1894, *Peneckiella* Soshkina, 1939, *Vestigiphyllum* Sytova *in* Sytova and Ulitina 1970, and *Fromeophyllum* Wright, 1981. Of these solitary and branching forms, *Macgeea*, *Thamnophyllum*, and *Peneckiella* are present in the Frasnian of western Canada, and are the subject of the following taxonomic study.

Genus *Macgeea* Webster, 1889

Pterorrhiza Ehrenberg, 1834, p. 312 (*nomen oblitum*).
Macgeea Webster, 1889, p. 710.
Pexiphyllum Walther, 1929, p. 128.
Trigonella Różkowska, 1980, p. 24 (*non* da Costa, 1778).
?*Debnikiella* Różkowska, 1980, p. 25.
Rozkowskaella Wrzołek, 1987, p. 277.

Type species. Pachyphyllum solitarium Hall and Whitfield, 1873.

Diagnosis (modified from Sorauf 1998, p. 85). Corallum solitary or with rare calicinal offsets, with prominent septal ridges above a low collar of epitheca in everted margin of calicinal area. Major and minor septa with symmetrical fans of rhipidacanthine trabeculae in dissepimentarium, where septa show varying degrees of fusiform dilation. Pipe of horseshoe dissepiments generally well developed, occurring at axis of divergence of trabecular fans. Stereome may coat horseshoes and adjacent septa, and may also fill early corallite growth stages. Small, globose dissepiments commonly occur interior to horseshoes, especially in late growth stages. Exterior to horseshoes, dissepiments may also be globose, but also may be flat or concave. Tabulae are most commonly flat or variably arched, with a broad axial platform.

Discussion. For many years there has been controversy over usage of the generic names *Macgeea* and *Pterorrhiza*. Neither of the authors of these names provided illustration or details of the internal morphology of their material (Webster 1889; Ehrenberg 1834). In the case of *Pterorrhiza*, Lang et al. (1940, p. 111) selected *Cyathophyllum marginatum* Goldfuss, 1826, from the Givetian of the Paffrather Mulde, Germany, as type species, while subsequently Birenheide (1969, p. 42; Pl. 3, Fig. 10; Pl. 5, Figs. 17a, b) selected a lectotype and illustrated it externally and in transverse section. However, neither the lectotype nor topotype material have been illustrated in longitudinal section and so the internal morphology of this species is inadequately known.

The type species of *Macgeea*, *Pachyphyllum solitarium* Hall and Whitfield, 1873, from the Cerro Gordo Member, Lime Creek Formation (late Frasnian) of Iowa, was illustrated externally and in diagrammatic transverse section by Hall and Whitfield (1873, Pl. 9, Figs. 6, 7). Sorauf (1998) selected a neotype for *M. solitaria* and provided a thorough study of morphology and variability in this species.

While *P. marginata* and *M. solitaria* are most probably congeneric, the better state of knowledge of the latter has led many authors to use *Macgeea* in preference to *Pterorrhiza*, despite the strict priority of *Pterorrhiza*. The case for suppression of *Pterorrhiza* as a *nomen oblitum*, initially argued by Schouppé and Cheng (1969), has been reviewed and amplified by Sorauf (1998, pp. 85–86). These arguments are not repeated here; interested readers are referred to these two publications and the references therein. Sorauf cited Article 79(c) of the 1985 edition of the International Code of Zoological Nomenclature in support of suppression of *Pterorrhiza* and the present author fully concurs in this view.

The genus *Pexiphyllum* was introduced by Walther (1929), although the type species, *P. rectum* Walther, 1929, was not designated until the work of Lang et al. (1940, p. 98). According to Birenheide and Liao (1985, pp. 249–250), Walther's material was from the Frasnian Flinz-Kalke of northwest Sauerland, Germany. Birenheide (1978) implied that *Pexiphyllum* could be separated from "*Pterorrhiza*" (= *Macgeea*, as interpreted here) by having globose outer dissepiments, as opposed to plate-like outer dissepiments in "*Pterorrhiza*". While there is a tendency for most species to have a preponderance of either globose or flat to concave outer dissepiments, many species show a mixture of both types, e.g., *M. solitaria* (Hall and Whitfield) as revised by Sorauf (1998), and Canadian forms such as *M. proteus* (Smith), *M. soraufi* n. sp., and *M. pustulosa* n. sp. As noted by McLean (1989, p. 245), the character of the dissepimentarium adjacent to the horseshoe pipe, although potentially having specific value, cannot be considered to be of generic significance, and *Pexiphyllum* is regarded here as a junior synonym of *Macgeea.*

Trigonella was erected by Różkowska (1980) with *T. sandaliformis* Różkowska, 1980 from the late Frasnian of the Holy Cross Mountains, Poland, as type and only species. It was based on a single, partly silicified specimen in which the outer dissepimentarium is lacking. Characteristic features were considered to be an everted dissepimentarium, lack of horseshoe dissepiments, and a calceoloid growth form (Różkowska 1980, p. 24). Wrzołek (1987) renamed the genus *Rozkowskaella*, since *Trigonella* is preoccupied by *Trigonella* da Costa, 1778, a modern bivalve. Coen-Aubert and Wrzołek (1991) restudied *R. sandaliformis* on the basis of new and more complete material from the type area and elsewhere, and produced a much clearer understanding of its morphology. They showed clear development of horseshoe dissepiments in two specimens (Pl. 1, Fig. 8; Pl. 2, Fig. 5) and regarded *Rozkowskaella* as a subgenus of *Macgeea*, differing from the latter only in lacking horseshoe dissepiments in late growth stages. However, of the three specimens illustrated in longitudinal section by Coen-Aubert and Wrzołek (1991), one (Pl. 1, Fig. 5) shows only the inner dissepimentarium, another (Pl. 2, Fig. 5) has horseshoes throughout, although it is not clear how close to the calice the section extends, while the third (Pl. 1, Fig. 8) shows both irregular horseshoes and well-developed ones in different parts of the corallite, but unfortunately is quite oblique in its upper part and it is not clear how the horseshoes are developed in that area. The longitudinal section of the holotype (Różkowska 1980, Fig. 4; Pl. 2, Fig. 6b) mostly shows only the inner dissepimentarium, with a few possible irregular horseshoes near the lower right corner. Therefore, based on the illustrated material, horseshoe development in this species is quite variable and their absence in late growth stages cannot be confirmed. Separation of this form from *Macgeea* at the subgeneric level does not seem justified.

Różkowska (1980) erected the genus *Debnikiella*, with type species *D. formosa* Różkowska, 1980, from the late Frasnian of the Silesian-Cracow Upland, Poland. It was based on a single, fragmentary specimen, which, on the only sectioned side of the corallum, has an arched dissepimentarium of small, globose forms with occasional broad horseshoes at its axis (Różkowska 1980, Fig. 5b; Pl. 4, Fig. 16b). Well-developed rhipidacanthine trabeculae are present within the dissepimentarium. The late Frasnian western Canadian species *Macgeea telopea* Crickmay and *M. pustulosa* n. sp., described herein, show a tendency to have a similar irregular development of horseshoe dissepiments in some specimens (or portions of some specimens), although normal horseshoes are also present in these species. Lack of knowledge of the variability within *D. formosa* precludes definite assignment of the species, but it seems probable, based at least on the variability within Canadian representatives of *Macgeea*, that it could be placed in that genus.

Coen-Aubert and Wrzołek (1991, p. 10) considered *Debnikiella* as a possible junior synonym of *Macgeea* (*Rozkowskaella*) Wrzołek. Later in that publication (p. 12), in redescribing *D. formosa*, they assigned it to *M.* (*Rozkowskaella*) without question, although in discussion of that species (p. 13), merely stated that "*Rozkowskaella* and *Debnikiella* are likely to be synonyms". While in the present interpretation *Rozkowskaella* is placed in synonymy with *Macgeea* sensu stricto and *Debnikiella* is regarded as a probable synonym of the latter, if *Rozkowskaella* and *Debnikiella* should be considered synonymous, but distinct from *Macgeea* (implied, but not totally clear in Coen-Aubert and Wrzołek 1991), then the appropriate name would be *Debnikiella* Różkowska, 1980, as it has priority over *Rozkowskaella* Wrzołek, 1987 (see International Commission on Zoological Nomenclature 1999, Articles 60.2, 60.3).

The genus *Protomacgeea* Różkowska, 1956 is of uncertain affinities, but needs to be considered in relation to *Macgeea*, especially as Różkowska (1957, p. 108) considered it to be probably ancestral to the latter genus. The type species, *P. dobruchnensis* Różkowska, 1956, from the early Eifelian (*patulus* zone, Fedorowski 2003, p. 107) of the Holy Cross Mountains, Poland, is a small, conical form with strongly dilated septa and flat outer dissepiments, reminiscent of some species of *Macgeea*, although lacking the everted calice characteristic of that genus (Różkowska 1956, Figs. 2–8). According to Różkowska (1956, p. 280) and Fedorowski (2003, p. 107), horseshoe dissepiments are lacking, and while trabecular fans do occur (Fedorowski 2003, p. 107), the presence of rhipidacanths has not been confirmed. Some Emsian forms from Australia have been assigned to *Protomacgeea* (see review by Zhen 1995), but the only one described in detail (*P. minor* Zhen, 1995 from the Mount Podge Limestone, Burdekin Basin, Queensland), has rhipidacanthine trabeculae and horseshoe dissepiments, together with a distinct axial columella, and is not congeneric with *P. dobruchnensis*. At present, it seems that *P. dobruchnensis* is unlikely to be a phillipsastreid and hence unrelated to *Macgeea*, but its affinities remain unclear. Even if shown to be a phillipsastreid, Różkowska's (1957) theory of *Protomacgeea* being ancestral to *Macgeea* cannot be supported, since older (Pragian and Emsian) species of

Macgeea are now recognized (see list of species included in *Macgeea*, below).

A number of species have been assigned in the past to *Macgeea* (or genera here considered synonymous with *Macgeea*), but are accommodated better elsewhere. *Cyathophyllum heterophylloides* Frech, 1885 (late Frasnian, Ibergerkalk, Harz Mts., Germany) was revised by Birenheide (1978) on the basis of Frech's illustrations and material from the Frasnian Refrath Formation, Bergisches Land, Germany. While Birenheide's material, which he assigned to *Pexiphyllum*, is clearly representative of *Macgeea*, as defined here (Birenheide 1978, Pl. 18, Figs. 4a–c), it is by no means certain that it is conspecific with any of Frech's material. As Birenheide noted (p. 113), it is certainly likely that Frech's material may include various species and genera. The transverse section (Frech 1885, Pl. 1, Fig. 2) upon which Birenheide based his definition of "*Pexiphyllum*" *heterophylloides*, has peripheral septal breakdown comparable to that of the charactophyllid genus *Piceaphyllum* Różkowska, 1980. The illustrated longitudinal section (Frech 1885, Pl. 1, Fig. 2), presumably from a different specimen, appears to lack horseshoe dissepiments, has some tendency to eversion of the dissepimentarium, and coarse trabeculae; it also may be a charactophyllid. Clearly Frech's material needs to be restudied if it is available, and a lectotype chosen.

Pexiphyllum altum Walther, 1929 (Frasnian, Flinz-Kalk, Sauerland, Germany) has well-developed charactophyllid trabeculae (author's examination of the holotype, Senckenberg Museum, Frankfurt) and, as stated by Birenheide (1978, p. 112), is a representative of *Temnophyllum* Walther, 1929.

Macgeea originata Soshkina, 1939 (Frasnian, Pashiya Horizon, western slopes of the southern Urals) is based on a single specimen, which lacks dissepiments. It was assigned to the siphonophrentid *Briantelasma* Oliver, 1960 by Ivanovskiy and Shurygina (1980), who provided photographic illustrations of the holotype for the first time. While it has the pronounced early septal dilation and cuneate late septal dilation typical of siphonophrentids such as *Enallophrentis* Oliver, 1993, it has quite coarse, low-angle trabeculae (Ivanovskiy and Shurygina 1980, Pl. 6, Fig. 1v) of apparent charactophyllid style, in contrast to the very fine trabeculae characteristic of the Siphonophrentidae (Oliver 1993). Its affinities are not clear, but it is certainly not a representative of *Macgeea*.

Stumm (1940) erected two species from the Frasnian Devil's Gate Limestone of Nevada, *Macgeea magna* from Devil's Gate and *M. subcylindrica* from Treasure Peak. The type material of *M. magna* in the U.S. National Museum consists of two incomplete transverse sections and a very incomplete longitudinal section, and all are labeled as "holotype". One of the transverse sections and the longitudinal section are those illustrated by Stumm (1940, Pl. 8, Figs. 8b, c) and stated by him (explanation of Plate 8, p. 65) to be from two different paratypes. The other transverse section illustrated by Stumm (Pl. 8, Fig. 8a), and stated to be a further paratype, is not the other section in the collection, and appears to be missing. The specimen illustrated in external view as holotype by Stumm (1940, Pl. 7, Fig. 5a) was apparently not sectioned and also seems to be missing. From that illustration it is abraded, but may be a representative of *Macgeea*. The paratypes illustrated in transverse section by Stumm may belong to *Macgeea*, but the longitudinal section is so incomplete that its assignment is not clear. If it is a *Macgeea*, then only some internal dissepiments are preserved, with the horseshoe pipe and any outer dissepiments removed. Clearly, restudy of this material with a larger topotype collection is necessary for it to be confirmed as a *Macgeea* and to be a useable species. As for the other species erected by Stumm, *M. subcylindrica*, it is represented in the U.S. National Museum collections by two transverse and one longitudinal section of the holotype and a longitudinal section of a paratype. One of the holotype transverse sections and the paratype longitudinal section were illustrated by Stumm (1940, Pl. 8, Figs. 7a, b). The non-illustrated longitudinal section of the holotype shows numerous small, globose, inwardly sloping dissepiments and very well-developed coarse, charactophyllid trabeculae. The holotype is clearly a charactophyllid on the basis of this longitudinal section and *M. subcylindrica* is regarded here as a representative of *Temnophyllum* Walther.

Macgeea ? crassoseptatum Wang, 1948 (Givetian, Yunnan) is a short septate form illustrated only in transverse section (Wang 1948, Pl. 2, Fig. 25). No mention was made of horseshoe dissepiments and its affinities are unclear. *Macgeea kozlowskii* Różkowska, 1953 (Frasnian, Holy Cross Mountains, Poland) is a branching form and, as noted by Różkowska (1957) and in the present work, is a representative of *Thamnophyllum* Penecke. *Pexiphyllum frechi* Sun, 1958 (Frasnian, Hunan) is also a branching form, but lacks horseshoe dissepiments. It should probably be assigned to *Disphyllum* de Fromentel. The phaceloid *Macgeea langi* Sun, 1958 (Frasnian, Shetianqiao Formation, Hunan) was said to lack horseshoe dissepiments and they are certainly not apparent in the fragmentary, oblique, longitudinal section (Sun 1958, p. 21, Pl. 11, Fig. 2b). It was placed in *Sulcorphyllum* Pedder, 1964 by Jia (1977, p. 148), but that genus is cerioid and regarded as a synonym of *Trapezophyllum* Etheridge, 1899 by McLean (1989). The affinities of *M. langi* remain uncertain and it could be a representative of *Peneckiella* Soshkina (see below).

Macgeea pulchra Spasskiy, 1960 (Gerikhov Beds, Frasnian, Rudniy Altai) has very coarsely dilated septa and abundant stereome coating of the tabulae. Horseshoe dissepiments are not apparent, nor are rhipidacanthine trabeculae (Spasskiy 1960, Pl. 17, Figs. 1, 2; Pl. 18, Figs. 1–4; Pl. 20, Fig. 1). It shows some general similarities to "*M*". *originata* Soshkina, 1939, discussed above, and like that species its affinities are unclear. *Macgeea calostrota* Crickmay, 1962 (Cooking Lake Formation, early Frasnian, Alberta) is an aphroid to thamnasterioid form, with individual corallites extending above the main, massive part of the corallum in late ontogeny. It was revised and re-assigned to *Pachyphyllum* Edwards and Haime by McLean (1986).

The type series of *Macgeea touti* Pedder, 1966 includes material from the late Emsian Loomberah Limestone and Sulcor Limestone of northern New South Wales. The holotype and other material from the Loomberah Limestone appear to have a normal pipe of horseshoe dissepiments, together with outer dissepiments typical of *Macgeea* (Pedder 1966, p. 187, Figs. 9, 11). However, some of the paratype Sulcor material has subsidiary rows of horseshoes diverging from the main horseshoe pipe into the outer dissepimentarium (Pedder 1966, p. 187; Wright 1981, Text-fig. 10d). Comparable diverging rows of horseshoe dissepiments were used by Wright (1981) as a defining character of his new fasciculate phillipsastreid genus *Fromeophyllum* (type species *F. climax* Wright, 1981, Mt. Frome Limestone, Emsian – ? early Eifelian, central New South Wales). Wright (1981) considered *M. touti* to be a possible representative of *Fromeophyllum*, but on the basis of what is currently known, only the Sulcor material,with diverging rows of horseshoes, shows the internal morphology of *Fromeophyllum*. Pedder (1966) regarded *M. touti* to be a solitary form, but as *F. climax* is only weakly colonial, with some apparently solitary fragments (Wright 1981, Pl. 93), it is possible that the Sulcor paratypes of *M. touti* could be fragments of a *Fromeophyllum*. Nevertheless, as the holotype of *M. touti* does not appear to have diverging rows of horseshoes, the species seems best retained in *Macgeea* for the present, with the possibility that the type series of *M. touti* may contain more than one species and genus.

Finally, two species erected by Fan *in* He and Fan (1988) and assigned to *Pexiphyllum teletabulatum* and *P. sichuanense* (Guanwushan Formation, late Givetian, Sichuan) lack horseshoe dissepiments and may be charactophyllids.

The following species are considered to be representatives of *Macgeea*, as here interpreted. Many are inadequately known and may prove to be synonymous with further study. After the type species, they are listed in chronological order of publication, with data on type stratigraphic horizon and locality.

Pachyphyllum solitarium Hall and Whitfield, 1873. Type species. Cerro Gordo Member, Lime Creek Formation, late Frasnian, Iowa. Revised by Sorauf (1998), who selected a neotype. *Macgeea culmula* Webster, 1889, from the same horizon and location, is most probably synonymous.

Cyathophyllum marginatum Goldfuss, 1826. Type species of *Pterorrhiza* Ehrenberg, 1834 (see above). Givetian, Paffrather Mulde, Germany. Revised by Birenheide (1969, 1978).

Caryophyllia dubia de Blainville, 1830. Frasnian, "Bensberg", near Cologne, Germany. Revised by Lang and Smith (1935) and Brice and Rohart (1974).

Cyathophyllum bathycalyx Frech, 1886. Freilingen Formation, late Eifelian, Dollendorfer Mulde, Germany. Revised and neotype selected by Schröder (1996).

Macgeea parva Webster, 1889. Independence Shale, mid Frasnian, Iowa. Revised herein.

Cyathophyllum araxis Frech, 1900. "Calceola-Schichten", ?Eifelian, Armenia. Givetian material from Armenia assigned to this species was described by Soshkina (1952) and Sytova and Ulitina (1974). The type material does not seem to have been restudied.

Cyathophyllum (*Thamnophyllum*) *multizonatum* Reed, 1922. Frasnian, Chitral, northern Pakistan. Revised by Schouppé (1965), based on material from the type area.

Pexiphyllum rectum Walther, 1929. Flinz-Kalk, Frasnian, northwest Sauerland, Germany. Type species of *Pexiphyllum* Walther, 1929.

Pexiphyllum primum Walther, 1929. Flinz-Kalk, Frasnian, northwest Sauerland, Germany. Considered as a possible synonym of "*P*". *rectum* by Birenheide (1978).

Pexiphyllum ultimum Walther, 1929. Flinz-Kalk, Frasnian, northwest Sauerland, Germany.

Pexiphyllum arcuatum Walther, 1929. Flinz-Kalk, Frasnian, northwest Sauerland, Germany. Placed in synonymy with "*P*". *ultimum* by Birenheide (1978).

Macgeea gallica Lang and Smith, 1935. Ferques Formation, early Frasnian, Boulonnais, northern France. Revised by Brice and Rohart (1974).

Campophyllum cylindricum Yoh, 1937. Givetian, Guangxi.

Macgeea berdensis Soshkina, 1939. Kyn (?) Horizon, Frasnian, central Urals. Revised by Ivanovskiy and Shurygina (1980).

Macgeea breviconus Soshkina, 1939. Kyn (?) Horizon, Frasnian, central Urals. Occurs with *M. berdensis* and regarded as a juvenile form of that species by Ivanovskiy and Shurygina (1980).

Macgeea socialis Soshkina, 1939. Frasnian, southern Urals. Considered a synonym of *M. berdensis* by Ivanovskiy and Shurygina (1980).

Macgeea proteus Smith, 1945. Member C (?), Hay River Formation, mid Frasnian, southern District of Mackenzie. Revised herein.

Macgeea thomasi Stainbrook, 1946. Independence Shale, mid Frasnian, Iowa.

Pexiphyllum bunthi Taylor, 1951. Plymouth Limestone, Givetian, Devon.

Macgeea caucasica Soshkina, 1952. Frasnian, Armenia.

Macgeea czarnockii Różkowska, 1953. Early Frasnian, Holy Cross Mountains, Poland. Regarded as a synonym of "*Pexiphyllum*" *rectum* by Birenheide (1978).

Pexiphyllum siemiradzkii Różkowska, 1953. Mid Frasnian, Holy Cross Mountains, Poland. Considered a synonym of "*Pexiphyllum*" *ultimum* by Birenheide (1978).

Macgeea bathycalyx kasimiri Różkowska, 1956. Givetian, Holy Cross Mountains, Poland. Regarded as a synonym of "*Pterorrhiza*" *marginata* by Birenheide (1978).

Macgeea bathycalyx josephi Różkowska, 1956. Givetian, Holy Cross Mountains, Poland.

Macgeea bathycalyx regularis Różkowska, 1956. Givetian, Holy Cross Mountains, Poland. Regarded as a synonym of *M. bathycalyx josephi* by Schröder (1996).

Macgeea bathycalyx longiseptata Różkowska, 1956. ?Givetian, Holy Cross Mountains, Poland. Considered a synonym of *M. bathycalyx josephi* by Schröder (1996).

Macgeea bathycalyx amabilis Różkowska, 1956. Givetian, Holy Cross Mountains, Poland. Possibly synonymous with *M. bathycalyx josephi*, according to Schröder (1996).

Macgeea ponderosa Stumm, 1960. Rhinestreet Formation, mid Frasnian, New York. Revised by Sorauf (1987).

Macgeea telopea Crickmay, 1962. Unnamed limestone unit, Imperial Formation, late Frasnian, southern District of Mackenzie. Revised herein.

Macgeea desioi Schouppé, 1965. Frasnian, Chitral, Pakistan.

Macgeea touti Pedder, 1966. Loomberah Limestone, late Emsian, northern New South Wales. See also discussion, above.

Pterorrhiza czarnockii brevisepta Pickett, 1967. Ibergerkalk, Frasnian, Weilberg/Lahn, Germany.

Pterorrhiza densa Fedorowski, 1968. Early Givetian, Holy Cross Mountains, Poland.

Macgeea minima Brice, 1970. Frasnian, Ghoudjerak, Afghanistan.

Macgeea gallica gigantea Brice and Rohart, 1974. Ferques Formation, early Frasnian, Boulonnais, northern France.

Macgeea multitabulata Jia, 1977. Liujiang Formation, Frasnian, Guangxi.

Neopetrozium guangxiense Jia, 1977. Dongjiangling Formation, late Givetian, Guangxi.

Macgeea houershanensis Yu and Liao, 1978*b*. Longdongshui Member, Houershan Formation, Eifelian, Guizhou.

Macgeea dahekouensis Kong *in* Kong and Huang 1978. Dushan Formation, Givetian, Guizhou.

Macgeea ephippia He, 1978. Ganxi Formation, Lower Devonian, Sichuan.

Macgeea shiziyaensis He, 1978. Guanwushan Formation, late Givetian, Sichuan.

Trigonella sandaliformis Różkowska, 1980. Late Frasnian, Holy Cross Mountains, Poland. Type species of *Rozkowskaella* Wrzołek, 1987. Revised by Coen-Aubert and Wrzołek (1991).

Macgeea inculta Jin and He, 1982. Dongjiangling Formation, late Givetian, Guangxi.

Macgeea bathycalyx yunnanensis Song, 1982. Upper Heyuanzhai Formation, ?early Frasnian, Yunnan. Considered a synonym of *M. bathycalyx* by Wang (1994).

Macgeea symmetrica Song, 1982. Upper Heyuanzhai Formation, ?early Frasnian, Yunnan. Revised by Wang (1994).

Pexiphyllum shidiaense Song, 1982. Upper Heyuanzhai Formation, ?early Frasnian, Yunnan. Revised by Wang (1994).

Macgeea gallica pauciseptata Coen-Aubert, 1982. Upper Aisemont Formation, late Frasnian, Belgium.

Macgeea rozkowskae Coen-Aubert, 1982. Lower Lustin Formation, early Frasnian, Belgium.

Macgeea lacroixi Coen-Aubert, 1982. Reefal member, Lustin Formation, mid Frasnian, Belgium.

Pexiphyllum multum Yu and Kuang, 1983. Mintang Formation, Givetian, Guangxi.

Pexiphyllum crassum Yu and Kuang, 1983. Mintang Formation, Givetian, Guangxi.

Pexiphyllum guangxiense Yu and Kuang, 1986. Late Givetian or early Frasnian, Guangxi.

Macgeea multiseptata Wang, 1994. Upper Heyuanzhai Formation, ?early Frasnian, Yunnan.

Macgeea yunnanensis Wang, 1994. Upper Heyuanzhai Formation, ?early Frasnian, Yunnan.

Pterorrhiza eifeliana Aung, 1995. Lower Maymyo Formation, Eifelian, northern Myanmar.

Macgeea camplannulata Sorauf, 1998. Owen Member, Lime Creek Formation, late Frasnian, Iowa.

Macgeea concinnula Sorauf, 1998. Nora Member, Shellrock Formation, mid Frasnian, Iowa.

Macgeea soraufi n. sp. Member C, Hay River Formation, mid Frasnian, southern District of Mackenzie.

Macgeea pustulosa n. sp. Jean-Marie Member, Redknife Formation, late Frasnian, southern District of Mackenzie.

Macgeea parva Webster, 1889
Pl. 1, Figs. 1–10

1889 *Macgeea parva* Webster, p. 711.

1946 *Macgeea parva* Webster; Stainbrook, p. 421, Pl. 57, Fig. 8; Pl. 58, Figs. 8, 9; Pl. 60, Figs. 6, 9.

1998 *Macgeea parva* Webster; McLean and Klapper, Pl. 2, Figs. 8, 9.

Material. GSC 126350; Ireton Formation at 3311.0 feet (1009.1 m), Imperial Egremont no. 109 well, Lsd 4, Sec. 18, Twp 58, Rge 21 W4M, Alberta (Well loc. 9 of Fig. 15). GSC 126351; Ireton Formation at 3311.3 feet (1009.2 m) in same well as GSC 126350, above. GSC 126352; Leduc Formation at 3313.5 feet (1009.9 m) in same well as GSC 126350, above. GSC 126353; Ireton Formation at 3157–3160 feet (962.2–963.1 m), Imperial Opal no. 38 well, Lsd 11, Sec. 20, Twp 58, Rge 22 W4M, Alberta (Well loc. 8 of Fig. 15).

Population studied. 4 sectioned coralla.

Range. Mid Frasnian, upper zone 11, *P. irregularis* coral fauna.

Diagnosis. Ceratoid to subcylindrical *Macgeea* characterized by very small corallum, having diameter 6–9 mm and 21–26 major septa. Major septa are long and typically twisted at the corallite axis. Horseshoe dissepiments are very small, outer dissepiments are flat, and inner dissepiments are rarely developed. Tabulae are relatively widely spaced.

Description. Corallum is ceratoid to subcylindrical, with height up to 18 mm, although more commonly 10–15 mm. Corallite diameter at base of calice is 6–9 mm, with a mean of 8.4 mm. Calice is typically deep, in largest specimen (GSC 126352, Pl. 1, Fig. 5) occupying approximately half of corallum height, and is generally flat based.

Major septa are long, generally slender axially near the calice, where they characteristically form a moderately developed vortex. They show some dilation and stereome development around the pipe of horseshoe dissepiments, and there is variable dilation in early growth stages. Minor septa range up to half length of major septa, which vary in number from 21 to 26 in late growth stages (mean 23).

Horseshoe dissepiments form a continuous pipe, are typically very small and are commonly coated with stereome, especially in early growth stages. Outer dissepiments, where preserved, are generally flat and in a single row. Inner dissepiments are rare and primarily developed above the level of the base of the calice. Tabulae are complete and incomplete, well spaced (0.5–1.5 mm), in some cases sagging peripherally and arched axially, while in others forming relatively flat series. Unevenly developed stereome commonly coats tabulae.

Remarks. M. parva was originally described (but not illustrated) by Webster (1889) from "Blue shales below the Devonian limestone" at Independence, Iowa. Subsequently, the species was revised by Stainbrook (1946), based on material he collected from the Independence Shale at the type locality and at Brandon, Iowa, and he noted (p. 422) that Webster's types were lost. The nature of the Independence Shale outcrops was controversial for many years, but it is regarded now as a "stratigraphic leak" of younger strata into sinkholes and caverns developed in the underlying late Middle Devonian Cedar Valley Formation (Klapper 1975). Klapper noted that the Independence was a mixture of a green-grey shale-bearing Frasnian conodonts, corals, and brachiopods, and a dark grey shale containing Carboniferous spores. As discussed by McLean and Sorauf (1989, p. 388), the coral fauna described by Stainbrook contains some characteristic Frasnian forms, but others that are not typically Frasnian, and it is quite possible that a mixture of faunas is present. Klapper and Lane (1985, p. 928) recognized the conodont *Palmatolepis semichatovae* Ovnatanova in the Independence and this species is also common in western Canada, characteristic of Montagne Noire Zone 11 (McLean and Klapper 1998, p. 518). It occurs associated with *Macgeea parva* in the lower Ireton Formation overlying the Leduc Formation of the Redwater reef (4–18–58–21W4M well, McLean and Klapper 1998, p. 547) in the Alberta subsurface. Other representatives of *M. parva* in Alberta are from the same stratigraphic level and thus the Iowa and Alberta occurrences are of comparable age.

The western Canadian material assigned here to *M. parva* shows no significant differences from that described by Webster (1889), and described and illustrated by Stainbrook (1946). Some of Stainbrook's material, newly sectioned, is illustrated here for comparison (Pl. 1, Figs. 1–3, 6). *M. parva* is an unusually small representative of the genus, and the only other species to show comparable size and other similarities are *M. minima* Brice, 1970 and *M. inculta* Jin and He, 1982. *M. minima* was described from the Frasnian of the Dacht-e-Nawar area, Afghanistan (Brice 1970; Pl.

15, Figs. 6a, b; Fig. 53). While having general similarities to the smallest specimens of *M. parva*, *M. minima* has very short major septa in the only known specimen. Lacking any data on its range of variability, *M. minima* can be separated from *M. parva* for now on the basis of that character. *M. inculta* is from the late Givetian Dongjiangling Formation of Guangxi, China and thus considerably older than *M. parva*. It has comparable dimensions to *M. parva*, but may be clearly distinguished by having much larger horseshoe dissepiments (Jin and He 1982, Pl. 30, Figs. 4a–c).

Macgeea proteus Smith, 1945
Pl. 1, Figs. 11–16; Pl. 2, Figs. 1–9; Pl. 3, Fig. 2

part. 1891 *Cyathophyllum caespitosum* Goldfuss; Whiteaves, p. 200, Hay River specimens only (see below), Pl. 27, Figs. 7, 8 (*non* Goldfuss 1826).
1901 *Cyathophyllum ceratites* Goldfuss; Lambe, p. 146 (*non* Goldfuss, 1826).
part. 1945 *Macgeea proteus* Smith, p. 27, Pl. 24, Figs. 2–6, 8–15 (not Pl. 24, Fig. 7, = *Macgeea pustulosa* n. sp.).
1956 *Macgeea proteus* Smith; Ma, p. 23.
1962 *Macgeea proteus* Smith; McLaren et al., p. 10, Pl. 4, Figs. 9, 10.
1970 *Pterorrhiza proteus* (Smith); McLaren et al., p. 620, Pl. 10, Figs. 29, 30.
1984 *Macgeea proteus* Smith; McLean, Pl. 1, Figs. 7, 8.

Type material. GSC 9300 (holotype); ? member C, Hay River Formation; Hay River, District of Mackenzie, locality 14 (Loc. 1c of Smith 1945, p. 67) of Locality Register and Fig. 5. GSC 9300a–c, e, j, l–n, p (paratypes); same horizon and location as holotype. GSC 9301 (paratype); ? member B or C, Hay River Formation; Hay River, District of Mackenzie, locality 14 (Loc. 1a of Smith 1945, p. 76) of Locality Register and Fig. 5. GSC 9302 (paratype); ? member B or C, Hay River Formation; Hay River, District of Mackenzie, locality 14 (Loc. 1b of Smith 1945, p. 67) of Locality Register and Fig. 5.

Additional material. GSC 4204a; 4204b, c; 4204d, e (these numbers represent three specimens, each now comprising two individually designated polished halves); ? member B or C, Hay River Formation; Hay River, District of Mackenzie, precise location unknown, see locality 75 (uncatalogued) of Locality Register and Fig. 5. GSC 126354–126360; member B, Hay River Formation, 1.95–22.65 m above base; left bank of Hay River, District of Mackenzie, locality 100 (Amoco Loc. 11131) of Locality Register and Fig. 5.

Population studied. 44 sectioned coralla.

Range. Mid Frasnian, zone 6 – ?early zone 11, *A. parvulum – Temnophyllum* sp. A coral faunas.

Diagnosis. Ceratoid to cylindrical *Macgeea*, with occasional lateral and calical offsets. Corallite diameter ranges from 15 to 24 mm, with 23–40 major septa generally extending 0.6–0.8 of corallite radius. Septa typically show moderate to strong fusiform dilation peripherally, becoming very slender and often contorted in tabularium. Pipe of horseshoe dissepiments well developed, but with very variable size of dissepiments in pipe, which is commonly coated in stereome. Outer dissepiments are flat, concave, or convex; inner dissepiments are numerous and globose to elongate. Tabulae are closely spaced, generally in arched series.

Description. Corallum is ceratoid to cylindrical, with cylindrical forms reaching at least 70 mm in height and being either straight or, more commonly, curved. Coralla are usually exclusively solitary, but may develop calicinal or lateral offsets, which, as noted by Smith (1945, p. 27), do not generally develop beyond an early stage. Largest offsets observed are calicinal and reach a height of 32 mm (GSC 126360, Pl. 2, Fig. 8). Calice is generally deep, with steep walls and flat or slightly arched base. Corallite diameter ranges from 15 to 24 mm in mature specimens, with a mean of 18 mm.

Major septa are typically somewhat withdrawn from corallite axis, extending 0.6–0.8 of corallite radius, and ranging in number from 23 to 40 in mature specimens (mean 32). They characteristically show strong fusiform dilation peripherally, becoming very slender and commonly contorted near their axial ends. Peripherally they are typically coated with stereome, but rhipidacanthine trabeculae may protrude to give a serrated appearance in transverse section. Minor septa also show fusiform dilation, but are generally significantly more slender than major septa, and usually barely extend axially beyond the inner margin of the dissepimentarium.

Horseshoe dissepiments form a continuous pipe and show highly variable coating with stereome. In specimens like the holotype (GSC 9300; Pl. 1, Figs. 11, 12) it coats the horseshoes and accessory dissepiments also, but in others (e.g., GSC 126359; Pl. 1, Figs. 13–15) it is virtually absent. Horseshoes are very variable in size and shape, often within the one specimen, but are generally small. Outer accessory dissepiments are generally well developed, very variable in shape and size, being convex, flat, or concave. In one example (GSC 126356, Pl. 2, Figs. 1, 2) a large protrusion on one side of the corallum contains abundant, small, globose, and elongate outer dissepiments; it presumably formed some sort of holdfast or support structure. Inner dissepiments are also highly variable in size, globose and elongate, and may form at least six rows. Tabulae are flat to moderately arched, generally with a broad, flattened central platform, complete and incomplete, and most commonly closely spaced (generally 1–4/mm). They are mostly not coated in stereome.

Remarks. M. proteus is a very common species in the upper Hay River Formation and, to a lesser extent, overlying Alexandra Member of the Twin Falls Formation of southern District of Mackenzie. It is also present in the Mikkwa Formation of northern Alberta, Perdrix Formation of the Alberta Rocky Mountains, and Leduc Formation of subsurface central Alberta. It was used to define a biostratigraphic zone within the upper Hay River Formation by Warren and Stelck (1950), but ranges beyond that level, as noted above and by McLean and Klapper (1998, p. 516).

M. proteus is closely comparable to *M. soraufi* n. sp., the latter being primarily distinguished by having abundant stereome development in early growth stages, and longer and more numerous major septa. Distinction between these two species is discussed further below in remarks on *M. soraufi*.

Several other species show similarities to *M. proteus*. *M. multizonata* (Reed 1922), originally described from the Frasnian of Chitral, northern Pakistan, and revised by Schouppé (1965), is a widespread Frasnian species, as discussed by Coen-Aubert (1982) and Rohart (1999) especially. Its similarities to *M. proteus* were noted by Smith (1945, p. 27), but it differs by being probably more consistently cylindrical, having generally larger and uniformly sized horseshoe dissepiments, slightly fewer septa, and generally more widely spaced tabulae.

M. solitaria (Hall and Whitfield 1873), from the late Frasnian Cerro Gordo Member of the Lime Creek Formation of Iowa, has been thoroughly revised by Sorauf (1998). It is also closely comparable to *M. proteus*, but specimens of similar size to the latter may be distinguished by having generally weaker development of internal accessory dissepiments, which results in the septa becoming more abruptly thinner axially beyond the horseshoe pipe, and thus less fusiform. *M. solitaria* also appears to lack offsets and has a greater tendency to be ovoid or angulate in transverse outline.

Also very similar to *M. proteus* is *M. gallica gallica* Lang and Smith, 1935, from the early Frasnian Ferques Formation of the Boulonnais, northern France (revised by Brice and Rohart 1974). It has a greater tendency to develop offsets than *M. proteus*, as noted by Smith (1945, p. 28), but also appears to differ in having less development of small, inner dissepiments, particularly as seen in the material of Brice and Rohart (1974).

Macgeea telopea Crickmay, 1962
Pl. 3, Figs. 1, 4, 6, 9; Pl. 4, Figs. 1–3, 9

1962 *Macgeea telopea* Crickmay, p. 4, Pl. 2, Figs. 12, 13; Pl. 4, Figs. 5–7.

Type material. PRI 27076 (holotype), PRI 27077, 27078 (paratypes); “Carlson Cr., NWT, 62°27′N, 123° 39′W. In Upper Devonian, Jean-Marie member of Grumbler Formation” (Crickmay 1962, p. 4). In current stratigraphic usage, this would indicate the unnamed limestone unit of the Imperial Formation. However, the co-ordinates given by Crickmay do not correspond to known outcrops of that unit (Douglas and Norris 1976), nor to the location of Carlson Creek (Fig. 3). Consequently, the precise collecting locality for the type suite remains uncertain. Subsequent collecting by the writer in the type horizon and general area mentioned by Crickmay failed to obtain more specimens of this species.

Additional material. GSC 126361; Mt. Hawk Formation, southeast face of Mt. Haultain, Alberta, locality 55 (R.K. Jull Loc. L-218) of Locality Register and Fig. 9.

Population studied. 4 sectioned coralla.

Range. Late Frasnian, zone 12, upper *M. pustulosa* – *H. magna* coral faunas.

Diagnosis. Subcylindrical to ceratoid *Macgeea*, characterized by large corallum with diameter 22–26 mm. Major septa range in number from 43 to 51, are moderately to weakly dilated peripherally and extend to 0.9 of corallite radius. Horseshoe dissepiments typically are very irregular in size, with large ones in places enclosing smaller ones. Inner and outer dissepiments are small, globose, and numerous. Tabulae are arched, with a broad, flat, central platform.

Description. Corallum is subcylindrical to ceratoid. Available material is fragmentary and represents only adult growth stages below the calice. According to the original description of Crickmay (1962, p. 4), coralla may reach a height of 90 mm, with a “shallow, bowl-shaped calice”. Corallite diameter ranges from 22 to 26 mm, with a mean of 24 mm. Cardinal fossula is well developed.

Major septa are characteristically long, extending 0.7–0.9 of corallite radius and ranging in number from 43 to 51 (mean 47). They show slight to moderate dilation in the dissepimentarium, where coarse rhipidacanthine trabeculae are developed, especially clearly seen in GSC 126361 (Pl. 3, Fig. 9). In the tabularium they are very slender and variably sinuous. Minor septa are confined to the dissepimentarium and are 0.3–0.5 of length of major septa. Stereome is not apparent in the available adult stage material.

Horseshoe dissepiments are highly variable in size and shape, with a large one commonly enclosing up to three smaller ones and forming a very irregular pipe (Pl. 3, Fig. 4; Pl. 4, Figs. 2, 9). Inner and outer dissepiments are abundant, small, mainly globose, and forming steeply inclined layers. Tabularium is broad, generally about 0.7 of corallite diameter, and typically comprising a wide central platform, sagging slightly axially and containing closely spaced complete and incomplete tabulae, surrounded by a narrow trough of sagging tabellae and incomplete tabulae. Stereome coating of tabulae or dissepiments is not apparent.

Remarks. Repository of the type material of *M. telopea* was not indicated in the original description of Crickmay (1962). Discovery of this material in the collections of the Paleontological Research Institution (P.R.I.), Ithaca, NY, showed that only some offcut fragments of the holotype and two paratypes (Crickmay 1962, Pl. 4, Figs. 5–7) were preserved, along with a series of acetate peels from those fragments. The transverse and longitudinal sections of the holotype illustrated by Crickmay (1962, Pl. 2, Figs. 12, 13) were absent. Through the courtesy of Dr. P. Hoover of the P.R.I., the writer was able to prepare new thin sections of the holotype and paratypes.

M. telopea is a rare species in western Canada, with only one additional specimen that seems clearly conspecific with the type material (GSC 126361, Pl. 3, Figs. 6, 9). Closest similarities lie with *M. pustulosa* n. sp., especially in the tendency to have irregularly sized horseshoe dissepiments, in places enclosing smaller ones. In fact, one specimen provisionally assigned to *M. pustulosa* is regarded here as transitional between the two species (GSC 126378, Pl. 6, Figs. 1, 2, see below). In general, undoubted *M. telopea* seems separable from the coeval but more widespread *M. pustulosa* by the former having considerably larger dimensions, more numerous septa, and development of outer dissepiments (see also remarks under *M. pustulosa*, below).

The combination of large corallite size, numerous septa and complex dissepimentarium clearly separates *M. telopea* from most other described species of *Macgeea*. Closest similarities may be seen with *M. gallica gigantea* Brice and Rohart, 1974 from the early Frasnian Ferques Formation, Boulonnais, northern France. This form shows comparable corallite size, septal number and character, irregular horseshoe dissepiments, and nature of the tabularium (Brice and Rohart 1974, Pl. 7, Fig. 6; Pl. 8, Figs. 1–5; Pl. 9, Figs. 8, 9). The significant difference appears to be the presence of flat or concave outer dissepiments in *M. gallica gigantea*, as compared to the small, globose forms of *M. telopea*. The nature of the outer dissepiments can be a highly variable character in some species of *Macgeea*, as noted elsewhere in this work. However available material of *M. telopea* and *M. gallica gigantea* seems to show fairly consistent and different development of these dissepiments and so they are considered separate for the present. A specimen from the Frasnian Refrath Formation of the Bergisches Land, Germany, illustrated by Birenheide (1978, Pl. 18, Figs. 4b, c) appears to differ from *M. gallica gigantea* only in its smaller dimensions. Birenheide regarded his form as synonymous with *Cyathophyllum heterophylloides* Frech, 1885, which he placed in the genus *Pexiphyllum*. He noted, however, that Frech's original material, from the Frasnian Ibergerkalk of Grund, Harz Mountains, Germany, may include various species and genera. In this regard, Frech's illustrated transverse sections of "*P*". *heterophylloides* (1885, Pl. 1, Fig. 2, 2a) may actually be representative of a charactophyllid such as *Piceaphyllum* Różkowska, and the longitudinal section does not show any horseshoe dissepiments (Pl. 1, Fig. 2b), although they may be obscured by trabeculae. Generic assignment of *Cyathophyllum heterophylloides* must await revision of Frech's material and selection of a lectotype, ideally with study of a larger collection from the Ibergerkalk. If the species can then be assigned to *Macgeea* and Birenheide's form included, then based for the moment on the characteristics of the latter, it is quite likely that *M. gallica gigantea* could be a junior synonym.

Macgeea soraufi n. sp.
Pl. 3, Figs. 3, 5, 7–12; Pl. 4, Figs. 4, 5, 11; Pl. 5, Figs. 1, 2, 5, 6

? 1988 *Macgeea thomasi* Stainbrook; Sorauf, p. 172, Figs. 17.6, 19.1–19.7, 20.1, 20.2 (?*non* Stainbrook, 1946).

Derivation of name. In honour of J. E. Sorauf, for his contributions to the study of the genus *Macgeea*.

Type material. GSC 126362 (holotype), 126363–126365 (paratypes); member C, Hay River Formation, 7.0–12.0 m above base, 0.0–5.0 m below top of member; Hay River, District of Mackenzie, locality 13 (Amoco Loc. 11210) of Locality Register and Fig. 5. GSC 126366 (paratype); member B, Hay River Formation; isolated outcrop in gravel pit by Hay River Highway, District of Mackenzie, locality 110 (Amoco Loc. 14302) of Locality Register and Fig. 5. GSC 126367–126369 (paratypes); Alexandra Member, Twin Falls Formation, 0.4–4.3 m above base; McNallie Creek, District of Mackenzie, locality 99 (Amoco Loc. 11146) of Locality Register and Fig. 5.

Population studied. 25 sectioned coralla.

Range. Mid Frasnian, zone 6 — ?early zone 11, *A. parvulum* – *Temnophyllum* sp. A coral faunas.

Diagnosis. Ceratoid to trochoid *Macgeea* with corallite diameter 19–24 mm and 36–48 major septa. Stereome is abundant, especially in early growth stages. Cardinal fossula is generally well developed, major septa typically reach almost to axis, outer dissepiments are commonly flat, and tabulae are closely spaced, in mainly flat series.

Description. Corallum is ceratoid to trochoid, generally with a circular transverse outline, although more rarely slightly ovoid. Calice is deep, especially in juvenile specimens (e.g., GSC 126368, Pl. 3, Fig. 7), with steep sides and in some cases showing a well-developed cardinal fossula. Corallite diameter ranges from 19 to 24 mm (mean 21 mm) in mature individuals, and height may reach at least 40 mm.

Major septa are generally long, usually reaching, or almost reaching corallite axis, and range in number from 36 to 48 in mature specimens, with a mean of 39. Minor septa are short, generally not extending axially as far as inner edge of zone of inner accessory dissepiments. Septa are characteristically heavily coated with stereome in early growth stages; in later stages the stereome is absent, or confined to axial regions and variable coating of the horseshoe pipe. Septa show moderate fusiform dilation near corallite margin, with major septa becoming slender and often sinuous near axis.

Horseshoe dissepiments form a continuous pipe and are typically very small. Outer accessory dissepiments are well developed and variable in character, ranging from flat to concave and convex, more rarely globose. Combinations of these characters occur in the one corallite, but flattened forms appear the most common. Inner accessory dissepiments are highly variable in size, although generally larger than the outer forms, and are globose to elongate, in 1–4 steeply inclined series. Tabulae are complete and incomplete, in flat to slightly arched series, and mainly closely spaced. They are commonly obscured by the abundant stereome in early growth stages.

Remarks. *M. sorauft* shows close similarities to material from the mid-late Frasnian Sly Gap Formation of New Mexico assigned by Sorauf (1988) to *M. thomasi* Stainbrook, 1946. Stainbrook's type material is from the mid Frasnian Independence Shale of Iowa and tends to be lobate, a little smaller, and with fewer major septa than *M. sorauft*, but its internal characters have been inadequately illustrated. Sorauf's material from New Mexico includes both lobate and circular forms and has well-developed stereome comparable to that in *M. sorauft* (e.g., Sorauf 1988, Figs. 19.3–19.5). It also has similar corallite dimensions and septal number. The New Mexico material is not convincingly conspecific with the type material of *M. thomasi* from Iowa, and it is certainly possible that the former could belong to *M. sorauft*. While revision of the Independence Shale fauna is clearly necessary to establish the diagnostic features of *M. thomasi*, it seems that *M. sorauft* can be separated at least on the basis of corallite growth form and size.

M. sorauft is closely comparable to *M. proteus* Smith, 1945, which also occurs in the Hay River Formation and Alexandra Member of southern District of Mackenzie, although as yet the two species have not been found at the same locations. While *M. proteus* has trochoid to ceratoid coralla comparable to *M. sorauft*, there is a strong tendency for *M. proteus* to form cylindrical coralla, a character not observed in *M. sorauft*. Additionally, *M. sorauft* may be distinguished by its abundant stereome development, generally smaller horseshoe dissepiments, and longer and more numerous major septa.

M. ponderosa Stumm, 1960, revised by Sorauf (1987), is from the mid Frasnian Rhinestreet Formation of New York, and is similar to *M. sorauft* in its growth form and abundant stereome development. However, it appears to be generally a little larger, with simpler, more widely spaced tabulae that appear to be axially sagging, or at least in sagging series.

Macgeea pustulosa n. sp.
Pl. 4, Figs. 6–8, 10; Pl. 5, Figs. 3, 4, 7–13; Pl. 6, Figs. 3–7

part. 1945 *Macgeea proteus* Smith, p. 27, Pl. 24, Fig. 7 only.

1998 *Macgeea* sp. B. McLean and Klapper, Pl. 1, Figs. 5, 8.

Derivation of name. Latin, pustulosus = full of blisters, referring to the abundant development of outer dissepiments characteristic of this species.

Type material. GSC 126370 (holotype), 126371 (paratype); Jean-Marie Member, Redknife Formation, 0.0–2.5 m below top, Table Rock Rapids, Trout River, District of Mackenzie, locality 38 (Amoco Loc. 15683) of Locality Register and Fig. 4. GSC 126372 (paratype); Jean-Marie Member, Redknife Formation, 0.0–3.75 m above base, downstream from Table Rock Rapids, Trout River, District of Mackenzie, locality 109 (Amoco Loc. 15670) of Locality Register and Fig. 4. GSC 9304 (paratype); same horizon and probably same location (Loc. 13 of Smith 1945, p. 71) as GSC 126372, above. GSC 126373 (paratype); Jean-Marie Member, Redknife Formation, 9.3–12.0 m below top, 0.0–2.7 m above base, Birch River, District of Mackenzie, locality 95 (Amoco Loc. 11175) of Locality Register and Fig. 4. GSC 126374 (paratype); Upper Member, Redknife Formation, isolated 5 m thick outcrop, Bouvier River, District of Mackenzie, locality 39 (Amoco Loc. 11155) of Locality Register and Fig. 4. GSC 126375 (paratype); Mt. Hawk Formation, 42.4–45.5 m below top, Winnifred Pass, Alberta, locality 26 (Amoco Loc. 12541) of Locality Register and Fig. 8. GSC 126376 (paratype); Southesk Formation, 5.0–6.0 m below contact with upper Mt. Hawk Formation, ridge approximately 5 km west of junction of Wildhay River and Eagle's Nest Creek, Persimmon Range, Alberta, locality 104 (Amoco Loc. 15254) of Locality Register and Fig. 9. GSC 126377 (paratype); argillaceous limestones and shales equivalent to the lower Nisku Formation at 13129.8 feet (4001.8 m), H.B. Amoco et al. Pinto well, Lsd 13, Sec. 19, Twp 69, Rge 8 W6M, Alberta (Well loc. 1 of Fig. 15).

Additional specimen questionably assigned to *M. pustulosa* (transitional to *Macgeea telopea*): GSC 126378; Jean-Marie Member, Redknife Formation, same horizon and location as GSC 126370, above.

Population studied. 49 sectioned coralla.

Range. Late Frasnian, zone 12, *M. pustulosa – H. magna* coral faunas.

Diagnosis. Turbinate to trochoid species of *Macgeea* with corallite diameter 14–20 mm and commonly 30–36 major septa. Septa are typically withdrawn from corallite margin, leaving wide zone of presepiments in outer dissepimentarium. Axially, major septa are slightly withdrawn and commonly contorted. Septal dilation is generally weak and largely confined to dissepimentarium. Horseshoe dissepiments are usually large and variable in size and shape, with large ones in places enclosing smaller ones. Outer dissepiments are extremely abundant, mainly small and globose, but in some cases becoming large and concave where dissepimentarium is widest. Inner dissepiments are generally absent in early growth stages, small, globose, and very variably developed in late stages. Tabulae are in flat to arched, closely spaced series.

Description. Corallum is primarily turbinate to trochoid, but may rarely approach a patellate growth form. Calice is typically deep, with a broad, flat to slightly arched base. Corallum height ranges from 13 mm in sub-patellate form (GSC 126371, Pl. 5, Fig. 12), to at least 38 mm in more common trochoid to turbinate specimens. Corallite diameter ranges from 14 to 20 mm, with a mean of 17 mm. Fossula is generally not pronounced.

Major septa extend from 0.7 to 0.9 of corallite radius, and are very slender and variably contorted in axial region. They range in number from 24 to 37, most commonly 30–36, with a mean of 32. Minor septa do not extend axially beyond the inner dissepimentarium. In specimens which have the outer dissepimentarium well preserved, septa are interrupted by the outermost dissepiments, resulting in a well-defined zone of presepiments. Both orders of septa show weak to moderate dilation within dissepimentarium, where the ends of the trabeculae commonly protrude to give the appearance of carinae in transverse section. Stereome development is absent in early growth stages and rare within dissepimentarium in late stages.

Horseshoe dissepiments form a generally continuous pipe and are very variable in size and shape, in some cases with large forms enclosing smaller ones. Inner dissepiments are generally small, globose to moderately elongate, and occur in 1–5 rows in adult growth stages. In early stages they are rare or more commonly absent. Outer dissepiments are commonly abraded, but in more complete specimens form a very broad zone. Close to the horseshoe pipe they are generally steeply inclined, small, and globose like the inner dissepiments, but towards the corallite margin they are more commonly large, elongate, and may locally become concave. This latter, outermost group forms the presepiments noted above. Tabulae are generally incomplete, closely spaced (2–4/mm), and in flat to moderately arched series.

Remarks. As noted above, *M. pustulosa* is closely related to *M. telopea* Crickmay, 1962, from coeval strata in western Canada. One specimen is provisionally regarded as *M. pustulosa*, but shows morphological characteristics (large corallites, large horseshoes) transitional to *M. telopea* (GSC 126378, Pl. 6, Figs. 1, 2). In the present work, it is unfortunate that only a small collection of *M. telopea* was available for study. However, on the basis of current material, *M. pustulosa* may be separated by its smaller size, fewer septa, and more abundant outer dissepiments. There is no evidence in *M. telopea* of the presepiment development so characteristic of well-preserved representatives of *M. pustulosa*, but the four specimens here assigned to *M. telopea* are probably abraded. Until a continuum in size and septal number between *M. telopea* and *M. pustulosa* can be demonstrated, together with clear development of a complex, wide outer dissepimentarium, including presepiments, in forms of *telopea* size, then it seems preferable to separate the two morphological groups at the species level.

Among non-western Canadian representatives of *Macgeea*, closest similarities appear to lie with the type species *M. solitaria* (Hall and Whitfield 1873) from the late Frasnian Cerro Gordo Member of the Lime Creek Formation, Iowa. A large population of this species was described by Sorauf (1998), who gave a good summary of its variability. The two species are comparable in corallite size and septal number, development and character of septa, especially the axial contortion of the major septa, and nature of inner dissepiments and tabulae. Major differences lie in the character of the outer dissepimentarium — in *M. solitaria* it is generally narrower and composed mainly of flat or concave forms; less commonly they are globose. In addition, there is no apparent tendency to develop the outermost zone of presepiments characteristic of *M. pustulosa*, while the pipe of horseshoe dissepiments seems to be more uniformly developed in *M. solitaria*.

Genus *Thamnophyllum* Penecke, 1894

Fascicularia Dybowski, 1873, p. 336 (*non* Lamarck, 1816).
Thamnophyllum Penecke, 1894, p. 563.
Phacellophyllum Gürich, 1909, p. 102.
?*Senceliastraea* Tsien, 1968, p. 450 (*nomen nudum*)
Profascicularia Cotton, 1973, p. 162.
?*Xystriphylloides* Yu, Liao and Deng, 1974, p. 224.
?*Neopetrozium* Jia, 1977, p. 151.
Thamnophylloides Jin and He, 1982, p. 117.
Pseudopexiphyllum Hubmann, 1992, p. 160.

Type species. *Thamnophyllum stachei* Penecke, 1894.

Diagnosis. Corallum fasciculate with lateral and axial increase. Adjacent corallites may be intermittently in lateral contact. Major and minor septa generally show varying fusiform dilation, with maximum dilation in dissepimentarium. Trabeculae are rhipidacanthine, forming fans centered over well-developed pipe of horseshoe dissepiments. Small globose to elongate dissepiments may occur on axial side of horseshoe pipe. Dissepiments outside horseshoe pipe, if present, vary from flat to concave and convex. If convex, then outer dissepiments are generally elongate and in one or more rows, usually most abundant joining offset corallites. Tabulae tend to be mainly flat and often complete in short septate species, variably arched and incomplete in long septate forms.

Discussion. The early history of study of the widespread Pragian to Frasnian genus *Thamnophyllum* has been thoroughly reviewed and summarized by Scrutton (1968, pp. 258–260), particularly the reasons for placing *Phacellophyllum* Gürich, 1909 in synonymy. The present writer is in full agreement with Scrutton's views and it is unnecessary to repeat the discussion here. However, since Scrutton's work a number of genera with *Thamnophyllum*-like morphology have been introduced, and review of these taxa is appropriate.

Sytova *in* Sytova and Ulitina (1970) proposed the genus *Vestigiphyllum*, with *Thamnophyllum tabulatum* Bulvanker, 1958, as type species. Bulvanker's material, from the late Emsian Shanda Beds of the Kuznets Basin, differs from typical *Thamnophyllum* in having markedly mesa-shaped tabulae, the central, arched parts of which join the axial ends of the major septa to give an "inner wall" appearance in transverse section (Bulvanker 1958, Pl. 38, Figs. 1a, b). Sytova's material from the late Eifelian of Armenia assigned to this species has consistent mesa-shaped tabulae (Sytova and Ulitina 1970, Pl. 43, Fig. 1b), and according to Sytova (p. 117) the axial ends of the major septa are curved to form an "axial tube" in conjunction with the central, arched portions of the tabulae. This curving of the septal ends and their involvement in the axial structure is not clear in Bulvanker's illustrated material, nor is it apparent in Sytova's illustrated transverse section of an individual corallite (Sytova and Ulitina 1970, Pl. 43, Fig. 1a). In the latter, the major septa actually seem to extend axially slightly beyond the "inner wall" of the vertically joined edges of the central arch of the tabulae. Whether or not the axial ends of the major septa in *T. tabulatum* do deflect and form an axial structure with the tabulae, which seems doubtful on the basis of the illustrated Kuznets Basin and Armenian material, the distinctive mesa-shaped tabulae forming a variably developed "inner wall", appears sufficient to separate the species from *Thamnophyllum*. Consequently, *Vestigiphyllum* is retained here as a valid genus.

Two genera from early Emsian strata (Pedder 1998, p. 237) of Guangxi, China, are poorly known, but may be synonyms of *Thamnophyllum*. *Xystriphylloides* Yu, Liao and Deng, 1974 (type species *X. nobilis* Yu, Liao and Deng, 1974) is a branching form in which horseshoe dissepiments are recorded, but not apparent in the unclear illustrations (Yu et al. 1974, Pl. 104, Figs. 17–20). Similarly, *Neopetrozium* Jia, 1977, based on *N. zhongguoense* Jia, 1977, also has unclear illustrations (Jia 1977, Pl. 51, Figs. 4a–d). It was stated to have a row of horseshoe dissepiments (Jia 1977, p. 151) and considered to differ from *Thamnophyllum* only by having convex tabulae. As *N. zhongguoense* has long major septa, convex tabulae would be expected, and if a continuous horseshoe pipe is confirmed in this species, it is probably a representative of *Thamnophyllum*. Pedder (1998) suggested both *Xystriphylloides* and *Neopetrozium* were probably synonyms of *Ivdelephyllum* Spasskiy, 1971, a genus based on *Keriophylloides caespitosum* Vaganova, 1959, from the late Emsian of the eastern Urals. *I. caespitosum* has an everted dissepimentarium lacking a horseshoe pipe (Vaganova 1959, Pl. 36, Fig. 3v) and is unrelated to *Thamnophyllum*. *Xystriphylloides* and *Neopetrozium* require further study before their affinities can be determined.

Thamnophylloides Jin and He, 1982, with type species *T. sinensis* Jin and He, 1982 from the Emsian Dale Member, Sipai Formation of Guangxi, is a branching form with long minor septa, strong peripheral septal dilation, a well-developed horseshoe pipe, and septa and outer dissepiments that connect at adjoining corallites (Jin and He 1982, Pl. 29, Figs. 1, 2). *T. sinensis* seems clearly to be a representative of *Thamnophyllum*, and, as in a previous review (McLean 1989), *Thamnophylloides* is regarded as a junior synonym of that genus.

Hubmann (1992) proposed the genus *Pseudopexiphyllum*, with type species *Thamnophyllum supradevonicum* Penecke, 1904, from the Frasnian of the Antitaurus Mountains, Turkey. This species has well-developed horseshoe dissepiments, together with abundant inner and particularly outer dissepiments (Penecke 1904, Pl. 4, Fig. 1b; Hubmann 1992, Text-Fig. 8, Pl. 2, Figs. 3–6) and thus is somewhat atypical of many species of *Thamnophyllum*. However, in the large *Thamnophyllum* population from western Canada examined in the present study, some species show extensive development of outer dissepiments in some, although not necessarily all, coralla. Examples of this feature may be seen in *T. colemanense* (Pl. 7, Figs. 4, 7; Pl. 9, Figs. 1–4), *T. tructense* (Pl. 10, Figs. 8, 10; Pl. 11, Figs. 1, 2), and *T. cordense* (Pl. 13, Figs. 7, 8; Pl. 14, Figs. 2, 4). It seems that development of this character within the overall *Thamnophyllum* morphological range is too variable to use as a basis for generic separation, and consequently *Pseudopexiphyllum* is regarded here as a junior synonym of *Thamnophyllum*.

The primarily cerioid to subcerioid phillipsastreid *Smithicyathus* Różkowska, 1980, with its possible synonym *Paramixogonaria* Liao and Birenheide, 1985, was reviewed by McLean (1989, 1994*b*). As was noted in those studies, there is a tendency for some species of *Smithicyathus*, notably the type, *S. cinctus* (Smith 1945) from the late Frasnian of western Canada, to develop a partially phaceloid corallum. Portions of a corallum with that growth form may show the general characters of *Thamnophyllum* (e.g., McLean 1994*b*, Pl. 2, Figs. 3, 5) and thus be confused with that genus. However, the preponderance of massive growth form in coralla of such species generally clearly separates them from *Thamnophyllum*. One species from the western Canadian Frasnian, *Thamnophyllum tructense* (McLaren 1959), revised below, may rarely show expansion of the outer dissepimentarium so that it approaches *Smithicyathus* in character (Pl. 11, Fig. 1). Although this species typically has well-developed outer dissepiments, its growth form is primarily phaceloid, with occasional lateral contact of corallites, and so is included here in *Thamnophyllum* without reservation (for further discussion, see remarks under *T. tructense*, below).

Several species originally assigned to *Thamnophyllum* lack horseshoe dissepiments and thus are specifically excluded from that genus. In this category are *T. curtum* Hill, 1941, from the Emsian Taemas Formation of the Wee Jasper area, New South Wales (revised and tentatively assigned to *Zelolasma* Pedder, 1964 by Pedder *in* Pedder et al. 1970), *T. abrogatum* Hill, 1941, from the Cavan Limestone of the same area (original illustrations of Hill 1941, Pl. 10, Figs. 4a, b inadequate for revised generic assignment), and *T. mirabile* Khoa *in* Khoa and Thom 1980, from the Givetian of Vietnam (revised by Thanh and Khoa 1988, who placed it in *Zelolasma*).

The following species are regarded here as representatives of *Thamnophyllum*. Many require revision to adequately understand their variability, and thus some may prove to be synonymous. After the type species they are listed in chronological order of publication, with data on type stratigraphic horizon and locality.

Thamnophyllum stachei Penecke, 1894. Type species. Barrandeikalk, Eifelian, Graz area, Austria. Revised by Lang and Smith (1935); lectotype illustrated by Flügel and Hubmann (1994, Pl. 2, Figs. 1, 2).

Lithodendron caespitosum Goldfuss, 1826. Type species of *Phacellophyllum* Gürich, 1909. ?Late Givetian, "Bensberg", Paffrather Mulde, Bergisch Gladbach, Germany. Holotype revised and illustrated by Lang and Smith (1935, Pl. 35, Figs. 1, 2) and Birenheide (1969, Pl. 2, Figs. 6a–f; Pl. 5, Fig. 15).

Cyathophyllum kunthi Dames, 1869 *sensu* Dybowski (1873). Type species of *Fascicularia* Dybowski, 1873 (*non* Lamarck, 1816) and *Profascicularia* Cotton, 1973. Late Frasnian, Sudeten Mountains, Poland. Reviewed by Coen-Aubert (1982). Not *C. kunthi* Dames *sensu* Różkowska (1957) = *Peneckiella*.

Diphyphyllum ?porteri Etheridge, 1890. Moore Creek Limestone, Eifelian, Tamworth area, New South Wales. Revised by Hill (1942*b*).

Thamnophyllum hoernesi Penecke, 1894. Barrandeikalk, Eifelian, Graz area, Austria. Revised by Lang and Smith (1935).

Thamnophyllum murchisoni Penecke, 1894. Barrandeikalk, Eifelian, Graz area, Austria. Revised by Lang and Smith (1935).

Diphyphyllum porteri var. *mitchellense* Etheridge, 1899. Kilgower Member, Tabberabbera Formation, ?late Pragian – early Emsian, Mitchell River area, Victoria. Revised by Talent (1963).

Thamnophyllum supradevonicum Penecke, 1904. Type species of *Pseudopexiphyllum* Hubmann, 1992. Frasnian, Antitaurus Mountains, Turkey. Revised by Hubmann (1992).

Thamnophyllum carnicum Vinassa de Regny, 1919. Middle Devonian, Val di Collina, Carnic Alps, Italy. Considered a synonym of *T. hoernesi* by Schouppé (1949). Revised by Assereto (1962).

Diphyphyllum colemanense Warren, 1928. Mt. Hawk Formation, ?mid-late Frasnian, southern Alberta Rocky Mountains. Revised herein.

Thamnophyllum reclinatum Hill, 1939. Loyola Limestone, Pragian, Mansfield area, Victoria. Revised by Philip (1962). Regarded as a synonym of *T. mitchellense* by Talent (1963).

Peneckiella monozonata Soshkina, 1939. Late Frasnian, western slopes of southern Urals. Revised by Ivanovskiy and Shurygina (1980). Considered here as a synonym of *T. colemanense.*

Thamnophyllum morganense Soshkina, 1949. Frasnian, southern Urals. Possibly synonymous with *T. colemanense.*

Thamnophyllum uniense Soshkina, 1951. Biya Horizon, ?late Emsian-early Eifelian, northern Urals.

Thamnophyllum virgatum Soshkina, 1951. Frasnian, central Urals.

Synaptophyllum soshkinae Różkowska, 1953. Early Frasnian, Holy Cross Mountains, Poland.

Macgeea kozlowskii Różkowska, 1953. Early Frasnian, Holy Cross Mountains, Poland. Regarded as a subspecies of *T. germanicum* by Scrutton (1968).

Thamnophyllum pertinense Soshkina, 1954. Semiluk Beds, mid Frasnian, Russian Platform.

Thamnophyllum skalense Różkowska, 1956. Early Givetian, Holy Cross Mountains, Poland. Regarded as a subspecies of *T. germanicum* by Scrutton (1968).

Thamnophyllum trigemme pajchelae Różkowska, 1956. Givetian, Holy Cross Mountains, Poland. Regarded as a subspecies of *T. germanicum* by Scrutton (1968).

Thamnophyllum kozlowskii superius Różkowska (1957). Mid Frasnian, Holy Cross Mountains, Poland. Regarded as a subspecies of *T. germanicum* by Scrutton (1968).

Thamnophyllum voronense Ermakova, 1957. Upper Voronezh Horizon, late Frasnian, Russian Platform. Considered here to be a synonym of *T. colemanense.*

Thamnophyllum tchumyshense Bulvanker, 1958. Shanda Beds, late Emsian, Kuznets Basin.

Thamnophyllum rzonsnickajae Bulvanker, 1958. Shanda Beds, late Emsian, Kuznets Basin.

Phacellophyllum fenense McLaren, 1959. Upper Member, Twin Falls Formation, mid Frasnian, southern District of Mackenzie. Revised herein and considered synonymous with *T. colemanense.*

Phacellophyllum tructense McLaren, 1959. Kakisa Formation, late Frasnian, southern District of Mackenzie. Revised herein.

Thamnophyllum tomiense Ivaniya *in* Zheltonogova and Ivaniya 1961. Glubok Beds, late Frasnian, Kuznets Basin.

Thamnophyllum germanicum germanicum Scrutton, 1968. Givetian, Eifel region, Germany. Includes many forms previously assigned to "*T. trigemme* (Quenstedt)" of Penecke (1894). Holotype revised and illustrated by Coen-Aubert (1998).

Thamnophyllum germanicum schouppei Scrutton, 1968. Early Givetian, Devon.

Thamnophyllum caespitosum paucitabulatum Scrutton, 1968. Late Givetian, Devon.

Phacellophyllum occlusum Tsien, 1969. Hanonet Formation, ?early Givetian, Dinant Basin, Belgium. Revised by Coen-Aubert (1998).

Phacellophyllum latum Tsien, 1969. "Couvinian", Dinant Basin, Belgium.

Phacellophyllum crustatum Tsien, 1969. Jemelle Formation (*fide* Coen-Aubert 1998), late Eifelian, Dinant Basin, Belgium.

Phacellophyllum gradatum Tsien, 1969. Hanonet Formation, ?early Givetian, Dinant Basin, Belgium.

Phacellophyllum kimberleyense Hill and Jell, 1970. Napier Formation, Frasnian, Canning Basin, Western Australia.

Phacellophyllum furcosum Pedder *in* Pedder et al. 1970. Taemas Formation, Emsian, Wee Jasper area, New South Wales.

Phacellophyllum furcosum var. *maius* Pedder *in* Pedder et al. 1970. Taemas Formation, Emsian, Wee Jasper area, New South Wales.

Thamnophyllum faveolatum Spasskiy *in* Dubatolov and Spasskiy 1971. Lower Devonian, Dzhungar Alatau.

Thamnophyllum sinense Yu, Liao and Deng, 1974. Dushan Formation, Givetian, Guizhou.

Thamnophyllum proprium Shurygina, 1977. Vizhay Horizon, late Pragian, eastern slopes of Urals.

Thamnophyllum guangxiense Jia, 1977. Dongjiangling Formation, late Givetian, Guangxi.

Phacellophyllum complicatum Kong *in* Kong and Huang 1978. Guanziyao Formation, Emsian, Guizhou.

Phacellophyllum crassum Kong *in* Kong and Huang 1978. Dushan Formation, Givetian, Guizhou.

Phacellophyllum houershanense Yu and Liao, 1978*b*. Longdongshui Member, Houershan Formation, Eifelian, Guizhou.

Thamnophyllum latum He, 1978. Guanwushan Formation, Givetian, Sichuan.

Phacellophyllum sichuanense He, 1978. Ganxi Formation, Emsian, Sichuan.

Phacellophyllum gaofangense Kong, 1981. Gaofengjie Formation, Emsian, Guangxi.

Thamnophyllum solutum Jin and He, 1982. Dale Member, Sipai Formation, Emsian, Guangxi.

Thamnophyllum tegulatum Jin and He, 1982. Dale Member, Sipai Formation, Emsian, Guangxi.

Thamnophylloides sinensis Jin and He, 1982. Type species of *Thamnophylloides* Jin and He, 1982. Dale Member, Sipai Formation, Emsian, Guangxi.

Neopetrozium gansuense Cao *in* Cao et al. 1982. Reer Group, Lower Devonian, Sichuan.

Phacellophyllum zhananense Ouyang *in* Cao et al. 1982. Lengshuihe Formation, Frasnian, Shaanxi.

Thamnophyllum hollardi Coen-Aubert, 1982. Late Frasnian, Belgium.

Thamnophyllum junggarense Cai *in* Cai and Zeng 1983. Kulumudi Formation, Middle Devonian, Xinjiang, NW China.

Thamnophyllum eminense Cai *in* Cai and Zeng 1983. Kulumudi Formation, Middle Devonian, Xinjiang, NW China.

Thamnophyllum qingheense Cai *in* Cai and Zeng 1983. Beitashan Formation, Middle Devonian, Xinjiang, NW China.

Thamnophyllum xiaoyigouense Liao and Li, 1991. Xihanshui Group, Givetian, Qinling Mountains, Gansu.

Thamnophyllum caespitosum leonense Birenheide and Soto, 1992. Portilla Formation, late Givetian, Léon, Spain.

Thamnophyllum germanicum moravicum Galle, 1993. Eifelian, Moravia.

Thamnophyllum turritum Coen-Aubert *in* Bertrand et al. 1993. L'Abîme Member, Couvin Formation, early Eifelian, Dinant Basin, Belgium.

Thamnophyllum simplex Errenst, 1993. Middle Givetian (Middle *varcus* subzone), northeast Sauerland, Germany.

Phacellophyllum padaukpinense Aung, 1995. Padaukpin biostrome, Eifelian, Padaukpin, northern Myanmar.

Thamnophyllum tsieni Coen-Aubert, 1998. Pondrôme Member, Jemelle Formation, early Givetian, Dinant Basin, Belgium.

Thamnophyllum cordense n. sp. Alexandra Member, Twin Falls Formation, mid Frasnian, southern District of Mackenzie.

Thamnophyllum pedderi n. sp. Upper Nisku Formation, late Frasnian, central Alberta.

Thamnophyllum julli n. sp. Zeta Lake Member, Nisku Formation, late Frasnian, central Alberta.

The following species may be representative of *Thamnophyllum*, but require further study and (or) clearer illustrations to confirm such a generic assignment:

Peneckiella achanayensis Soshkina, 1939. Late Frasnian, western slopes of southern Urals. See remarks below on *T. pedderi*.

Peneckiella spiralis Soshkina, 1939. Late Frasnian, western slopes of southern Urals. See remarks below on *T. colemanense*.

Macgeea (Thamnophyllum) peneckei Schouppé, 1949. ?Eifelian, Graz area, Austria.

Disphyllum (Phacellophyllum) varians Taylor, 1951. "Plymouth Limestone", Middle Devonian, Devon. Stated to be solitary, but if a fragment of a colonial form, would probably be a representative of *Thamnophyllum*.

Phacellophyllum pseudotrigemme Tsien 1969. "Couvinian", Dinant Basin, Belgium.

Xystriphylloides nobilis Yu, Liao and Deng, 1974. Early Emsian, Guangxi. Type species of *Xystriphylloides* Yu, Liao and Deng, 1974. See discussion of *Thamnophyllum*, above.

Phacellophyllum ossalense Joseph and Tsien, 1975. Early Givetian, western Pyrénées, southern France.

Neopetrozium zhongguoense Jia, 1977. Yujiang Formation, early Emsian, Guangxi. Type species of *Neopetrozium* Jia, 1977. See discussion of *Thamnophyllum*, above.

Neopetrozium minor Jia, 1977. Yujiang Formation, early Emsian, Guangxi.

Phacellophyllum jiangzhaiense Kong *in* Kong and Huang 1978. Dushan Formation, Givetian, Guizhou.

Phacellophyllum alveolatum Yu and Liao, 1978*a*. Alengdu Formation, Lower Devonian, Yunnan.

Phacellophyllum daleense Yoh *in* Yoh and Bai 1978. Sipai Formation, Emsian, Guangxi.

Thamnophyllum dohmi Ivaniya, 1980. Shanda Beds, late Emsian, northeast Salair.

Phacellophyllum abnorme Kong, 1981. Gaofengjie Formation, Emsian, Guangxi.

Phacellophyllum longiseptatum Ouyang *in* Cao et al. 1982. Cakuohe Formation, Frasnian, Gansu. Possibly a representative of *Smithicyathus* Różkowska, as discussed by McLean (1994*b*).

Phacellophyllum xinhuaense Jin, 1982. Yutianqiao Stage, Frasnian, Hunan.

Phacellophyllum peneckielloides Ouyang *in* Cao and Ouyang 1987. Cakuohe Formation, Frasnian, Qinling Mountains, China.

Thamnophyllum khelopense Khoa, ? year of publication. Eifelian, Vietnam. Illustrated by Thanh and Khoa (1988, Pl. 57, Figs. 3, 4), but indeterminate on the basis of those illustrations. (?*nomen nudum*).

Thamnophyllum colemanense (Warren 1928)

Pl. 6, Figs. 8–15; Pl. 7, Figs. 1–7; Pl. 8, Figs. 1–9; Pl. 9, Figs. 1–7, 11

1928 *Diphyphyllum colemanense* Warren, p. 116, Pl. 1, Fig. 18.

? 1935 *Synaptophyllum arundinaceum* (Billings); Lang and Smith, p. 561, Text-Figs. 19, 20 (*non* Billings, 1859).

1939 *Peneckiella monozonata* Soshkina, p. 25, Pl. 10, Figs. 76–78.

? 1939 *Peneckiella spiralis* Soshkina, p. 28, Pl. 10, Figs. 79, 80.

part. 1945 *Disphyllum (Synaptophyllum)* cf. *arundinaceum* (Billings); Smith, p. 22, Pl. 12, Figs. 1, 2 (*non* Billings 1859). (GSC 9330a only, not GSC 9330 = *Peneckiella floydensis*).
? 1949 *Synaptophyllum arundinaceum* (Billings); Stumm, Pl. 17, Figs. 19, 20 (*non* Billings 1859).
? 1949 *Synaptophyllum arundinaceum* (Billings); Schouppé, Pl. 12, Figs. 60, 61 (*non* Billings 1859).
part. 1949 *Macgeea (Thamnophyllum) hoernesi* Penecke; Schouppé, p. 128, Pl. 11, Figs. 44–46 only (*non* Penecke 1894).
? 1949 *Macgeea (Synaptophyllum) spiralis* (Soshkina); Schouppé, p. 171, Pl. 12, Figs. 64, 65.
1949 *Thamnophyllum monozonatum* (Soshkina); Soshkina, p. 77, Pl. 32, Figs. 6, 7.
? 1949 *Thamnophyllum morganense* Soshkina, p. 80, Pl. 32, Figs. 8–10.
1951 *Thamnophyllum monozonatum* (Soshkina); Soshkina, p. 78, Pl. 13, Fig. 7; Text-Fig. 33.
1952 *Thamnophyllum monozonatum* (Soshkina); Soshkina, p. 85, Pl. 19, Fig. 71.
non 1953 *Thamnophyllum monozonatum* (Soshkina); Różkowska, p. 14, Text-Fig. 5; Pl. 1, Figs. 8–10 (= *T. kozlowskii superius* Różkowska, 1957).
? 1956 *Synaptophyllum arundinaceum* (Billings); Hill, Fig, 191–6a, b (*non* Billings 1859).
1957 *Thamnophyllum voronense* Ermakova, p. 168, Pl. 4, Figs. 1–3.
1959 *Phacellophyllum fenense* McLaren, p. 29.
1965 *Thamnophyllum monozonatum* (Soshkina); Ivaniya, p. 115, Pl. 93, Fig. 389; Pl. 94, Figs. 390, 391.
? 1974 *Thamnophyllum monozonatum* (Soshkina); Sytova and Ulitina, p. 31, Pl. 1, Figs. 1a, b.
1975 *Thamnophyllum monozonatum* (Soshkina); Besprozvannykh et al., p. 58, Pl. 12, Figs. 4, 5.
? 1978 *Phacellophyllum fenense* McLaren; Luke, p. 92, Pl. 2, Figs. 1, 2.
part. 1980 *Thamnophyllum monozonatum* (Soshkina); Ivanovskiy and Shurygina, p. 23.
1981 *Thamnophyllum monozonatum* (Soshkina); Tsyganko, p. 76, Pl. 28, Figs. 2, 3.
? 1993 *Thamnophyllum monozonatum* (Soshkina); Soroka, p. 20, Text-Fig. 3.
1998 *Thamnophyllum* sp. C. McLean and Klapper, p. 520, Pl. 1, Figs. 9, 10.

Type material. UA Dv 647 (lectotype, here designated); Mt. Hawk Formation, Crowsnest Pass, southern Alberta, locality 119, (third uncatalogued locality) of Locality Register and Fig. 14.

Additional material. GSC 9330a (holotype of *Phacellophyllum fenense* McLaren, 1959); upper member, Twin Falls Formation, Hay River near Grumbler Rapids, District of Mackenzie, locality 46 (Loc. 5 of Smith 1945, p. 67) of Locality Register and Fig. 5. GSC 126379; upper member, Twin Falls Formation, right bank of Hay River at Grumbler Rapids, District of Mackenzie, locality 47 (GSC Loc. 30442) of Locality Register and Fig. 5. GSC 126380–126382; upper member, Twin Falls Formation, right bank of Hay River, approximately 1.6 km downstream from Grumbler Rapids, District of Mackenzie, locality 45 (GSC Loc. 30444) of Locality Register and Fig. 5. GSC 126383; Simla Formation, 0.0–5.0 m above base, eastern flank of Bastille Mountain, British Columbia, locality 71 (Amoco Loc. 15261) of Locality Register and Fig. 8. GSC 126384; Mt. Hawk Formation, 35.4 m below top, Winnifred Pass, Alberta, locality 26 (Amoco Loc. 12541) of Locality Register and Fig. 8. GSC 126385, 126386; upper Southesk Formation, 25.0–36.0 m below base of Simla Formation, eastern side of South Berland River, Persimmon Range, Alberta, locality 85 (Amoco Loc. 15253) of Locality Register and Fig. 9. GSC 126387; Southesk Formation, southeast face of Mt. Haultain, Alberta, locality 55 (R.K. Jull Loc. L-223) of Locality Register and Fig. 9. GSC 126388, possible topotype; Mt. Hawk Formation, approximately 5 m above base of resistant carbonates, Crowsnest Pass, Alberta, locality 119 (first uncatalogued locality) of Locality Register and Fig. 14. GSC 15517, 15518, possible topotypes; Mt. Hawk Formation, 173.7–176.8 m below base of Palliser Formation, Crowsnest Pass, Alberta, locality 119 (second uncatalogued locality) of Locality Register and Fig. 14. GSC 126389; upper Nisku Formation, at 8319.4 feet (2535.6 m), Arco Sturgeon Lake South well, Lsd 9, Sec. 27, Twp 70, Rge 23 W5M, Alberta (Well loc. 2 of Fig. 15). GSC 126390; upper Nisku Formation, at 7555.5 feet (2302.8 m), CPOG Wallace River well, Lsd 6, Sec. 3, Twp 67, Rge 15 W5M, Alberta (Well loc. 4 of Fig. 15). GSC 126391; Wolf Lake Member, Nisku Formation, at 5026 feet (1531.9 m), Texaco McColl Majeau Lake no. 1 well, Lsd 12, Sec. 1, Twp 57, Rge 3 W5M, Alberta (Well loc. 7 of Fig. 15). GSC 126392; Lobstick Member, Nisku Formation, at 2394.8 m, Amoco et al. Bigoray well, Lsd 4, Sec. 29, Twp 52, Rge 8 W5M, Alberta (Well loc. 12 of Fig. 15). GSC 126393; Wolf Lake Member, Nisku Formation, at 8299.5 feet (2529.6 m), HB Pembina well, Lsd 12, Sec. 16, Twp 50, Rge 9 W5M, Alberta (Well loc. 16 of Fig. 15).

Population studied. 120 sectioned coralla.

Range. Mid – late Frasnian, zones 11–12, *P. irregularis* – *H. magna* coral faunas.

Diagnosis. Phaceloid *Thamnophyllum* with corallite diameter 5.5–9 mm and 15–23 major septa. Major septa vary in length from 0.5 of corallite radius to almost reaching corallite axis. Septa usually show moderate peripheral dilation, but may rarely be undilated. Horseshoe dissepiments form a uniform pipe and are relatively large. Inner dissepiments are generally absent; outer dissepiments commonly occur at lateral offsets, and in well-preserved specimens are seen elsewhere as flat or slightly arched, downsloping plates. Tabulae are incomplete and typically form moderately to strongly arched series.

Description. Corallum is phaceloid, with very variably spaced corallites, and lateral, non-parricidal increase. Offsets are typically buttressed with outer dissepimental plates. Adult corallites range in diameter from 5.5 to 9 mm, with a mean of 7.4 mm.

Major septa vary in number from 17 to 23, with 19–20 being most common, and a mean of 19. They are quite variable in length, extending from about half the corallite radius, to more commonly 0.7–0.8 of that radius; more rarely they can almost reach the corallite axis. In the tabularium major septa are very slender and may show contortion in the axial region. Minor septa generally barely extend axially beyond the peripheral horseshoe pipe. Peripheral dilation of both sets of septa is highly variable, ranging from complete lack of dilation to, in rare cases, adjacent septa being almost in lateral contact. More typically, weak to moderate dilation occurs, with well-developed rhipidacanthine trabeculae. Stereome coating of peripheral parts of the septa and the horseshoe pipe is characteristic, except in corallites lacking peripheral septal dilation.

Horseshoe dissepiments are generally uniform in size in each corallum, relatively large, and form a continuous pipe. Inner dissepiments are usually absent, but if present they are globose and typically only intermittently developed in an individual corallite. Outer dissepiments are well developed at lateral corallite offsets and are generally small and elongate. Elsewhere they are usually absent (probably by abrasion), but if preserved they are planar or slightly curved, and downward sloping from the horseshoe pipe, in 1–2 series, rarely more.

Tabulae are incomplete and form variably arched series. Short septate corallites tend to have flatter series, while corallites with longer major septa typically have well-arched series, with peripheral troughs and more flattened axial parts. They are generally closely spaced, commonly 2–3/mm.

Remarks. Thin section illustrations of the type material of *Diphyphyllum colemanense* Warren, 1928, from the Mt. Hawk Formation at Crowsnest Pass, southern Alberta, are given here for the first time (Pl. 6, Figs. 12, 13) and a lectotype chosen. This specimen is regarded here as conspecific with the holotype of *Phacellophyllum fenense* McLaren, 1959, from beds now regarded as in the uppermost part of the Twin Falls Formation on Hay River, District of Mackenzie. The latter specimen (GSC 9330a, Pl. 8, Figs. 3–6, 8) was incorrectly referred to as GSC 9330 by both Smith (1945) and McLaren (1959). GSC 9330 is assigned to *Peneckiella floydensis* (see below). Material from the Grumbler Rapids area on Hay River (GSC 126379, Pl. 9, Figs. 1–4; GSC 126380, Pl. 6, Figs. 8, 11; GSC 126382, Pl. 6, Figs. 9, 10) is among the most complete and shows the greatest observed development of outer dissepiments, in some cases connecting adjoining corallites. Luke (1978) referred material from the Frasnian part of the Guilmette Formation of the Leppy Range, eastern Nevada, to *P. fenense*. Luke's illustrations (Pl. 2, Figs. 1, 2) are unclear and the stated range in corallite diameter (8–11 mm) and septal number (44–50) overlaps the high end observed for Canadian *T. colemanense*. The Nevada material may belong to *T. colemanense*, but further study is needed for confirmation.

Among Canadian species, *T. colemanense* shows closest similarities to T. *pedderi* n. sp. The latter may be distinguished by its typically smaller corallites, fewer and shorter major septa, smaller horseshoe dissepiments, and simpler tabulae. Some smaller, shorter septate forms of *T. colemanense* approach *T. pedderi* in character, but the two populations as a whole are clearly separable.

A number of Russian Frasnian representatives of *Thamnophyllum* appear closely comparable to *T. colemanense*. *T. monozonatum* (Soshkina 1939), from the late Frasnian of the western slopes of the southern Urals, has similar dimensions, septal number and development, and character of horizontal skeletal elements (Soshkina 1939, pp. 25–26, Pl. 10, Figs. 76–78; 1949, pp. 77–78, Pl. 32, Figs. 6, 7; 1951, pp. 78–79, Pl. 13, Fig. 7, Text-Fig. 33; 1952, p. 85, Pl. 19, Fig. 71). It is regarded here as a synonym of *T. colemanense*. Material from various other regions has been assigned to *T. monozonatum*. Although variability in their material was not illustrated, that of Ivaniya (1965, Glubok Beds, late Frasnian, Kuznets Basin, Pl. 93, Fig. 389; Pl. 94, Figs. 390, 391), Besprozvannykh et al. (1975, Frasnian, Sette-Daban Range, NE Russia, Pl. 12, Figs. 4, 5) and Tsyganko (1981, Frasnian, northern Urals, Sub-Arctic Urals, Pay-Khoy, Pl. 28, Figs. 2, 3) is likely to belong to *T. monozonatum* and hence *T. colemanense*. The material from the Frasnian of Armenia placed in *T. monozonatum* by Sytova and Ulitina (1974, Pl. 1, Fig. 1) is a little larger (corallite diameter 10–14 mm) than typical *T. colemanense*, and may have somewhat shorter major septa and more widely spaced tabulae, although the range of variation was not described. It is considered here as a possible synonym of *T. colemanense*. Soroka (1993) discussed the early blastogeny of material from the Frasnian Sirachoy Formation of southern Timan that was assigned to *T. monozonatum*. It shows well-developed outer dissepiments around lateral offsets, comparable to *T. colemanense* (Soroka 1993, Fig. 3), but as adult forms were neither described nor illustrated, it cannot be certain if this material should be assigned to *T. colemanense*.

Schouppé (1949) placed *T. monozonatum* in synonymy with *T. hoernesi* Penecke, 1894, an Eifelian species from the Barrandeikalk of the Graz area, Austria. However, *T. hoernesi* has much larger corallites (15–20 mm diameter), very small horseshoe dissepiments, and widely spaced, sagging tabulae (Penecke 1894, Pl. 7, Figs. 13, 14). It seems clearly unrelated to *T. monozonatum* and *T. colemanense*. Różkowska (1953) assigned some material from the mid Frasnian of the Holy Cross Mountains, Poland, to *T. monozonatum*. However, it has axial, parricidal increase, short, strongly dilated septa, small horseshoe dissepiments, and well-spaced, sagging tabulae (Różkowska 1953, Pl. 1, Figs. 8–10). It was subsequently placed in a new subspecies, *T. kozlowskii superius*, by Różkowska (1957).

Ivanovskiy and Shurygina (1980) regarded a number of species from the Frasnian of the southern Urals, described by Soshkina (1939–1951), as synonyms (or possible synonyms) of *T. monozonatum*, and hence they need to be considered in relation to *T. colemanense*. *Peneckiella achanayensis* Soshkina, 1939 may be a representative of *Peneckiella* or *Thamnophyllum*, as discussed below in remarks on *T. pedderi* n. sp., but if it can be assigned to *Thamnophyllum*, its smaller corallites and simpler horizontal skeletal elements would seem to separate it from *T. colemanense*. *Peneckiella nalivkini* Soshkina, 1939 has dimensions that fit in the range of the Canadian material of *T. pedderi* n. sp. (corallite diameter 4–4.5 mm, 15–17 major septa: Soshkina 1939, p. 24), and this is also considerably smaller than *T. colemanense*. While having some horseshoe dissepiments, *P. nalivkini* appears to have some "unrolling" of other dissepiments (Soshkina 1939, Pl. 8, Figs. 6, 9) and may be a true representative of *Peneckiella* (see discussion of that genus, below). *Peneckiella spiralis* Soshkina, 1939 was originally regarded as solitary, but Ivanovskiy and Shurygina (1980) suggested it could be a fragment of a colonial form (the original material was stated by them to be lost). If it is colonial, then it could be a synonym of *T. colemanense*; its dimensions certainly fit within the range of that species. *T. morganense* Soshkina, 1949 is comparable to *T. colemanense* in corallite dimensions, but appears to differ primarily in having very pronounced septal dilation, even within the tabularium (Soshkina 1949, Pl. 32, Figs. 8–10). However, its range of variation is unknown and it could be synonymous with *T. colemanense*. *T. virgatum* Soshkina, 1951 is a species with small corallites (diameter 3.5–5 mm) and major septa numbering 14–15, and on that basis alone seems separable from *T. colemanense*, although in other respects it appears comparable (Soshkina 1951, Pl. 13, Figs. 2–6). Finally, the material regarded as *Peneckiella minima* (Roemer) by Soshkina (1951), and placed in *T. monozonatum* by Ivanovskiy and Shurygina (1980), seems generically indeterminate based on the published illustrations (Soshkina 1951, Pl. 20, Figs. 2–4).

One other described species appears to fit within the range of variation of *T. colemanense*, although its own variability has not been determined. *T. voronense* Ermakova, 1957 (upper Voronezh Horizon, Frasnian, Russian Platform) has some corallites slightly larger than the range of *T. colemanense*, but otherwise shows no significant differences (Ermakova 1957, p. 168, Pl. 4, Figs. 1–3). It is included here in *T. colemanense*.

T. tomiense Ivaniya *in* Zheltonogova and Ivaniya 1961 (Glubok Beds, late Frasnian, Kuznets Basin) is closely similar to *T. colemanense*, differing only in having some corallites with major septa joining in groups at the axis to give an almost bilateral symmetry (Zheltonogova and Ivaniya 1961, Pl. D-56, Fig. 2a; Ivaniya 1965, Pl. 93, Fig. 387). While long septate forms of *T. colemanense* do occur, this approach to bilateral symmetry has not been observed and seems sufficient to separate *T. tomiense* from *T. colemanense*. *T. zhananense* (Ouyang *in* Cao et al. 1982) from the Frasnian Lengshuihe Formation of Shaanxi has comparable dimensions to *T. colemanense*, but seems to have greater septal dilation in the tabularium. The illustrations (Cao et al. 1982 Pl. 28, Figs. 4a, b) are too unclear for adequate characterization of the species and it seems preferable to retain its separate status for the present.

Thamnophyllum tructense (McLaren 1959)
Pl. 9, Figs. 8–10, 12, 13; Pl. 10, Figs. 1–10; Pl. 11, Figs. 1–4

part. 1945 *Disphyllum* (*Synaptophyllum*) *stramineum* (Billings); Smith, p. 23, Pl. 13, Figs. 1, 2, 4, 5, 7, 8, 10 only (*non* Billings 1859).
? 1952 *Thamnophyllum stramineum* (Billings); Soshkina, p. 85 (*non* Billings 1859)
1959 *Phacellophyllum tructense* McLaren, p. 29.
non 1960 *Thamnophyllum stramineum* (Billings); Spasskiy, p. 48, Pl. 15, Figs. 1–3 (*non* Billings 1859).
1962 *Thamnophyllum tructense* (McLaren); McLaren et al., p. 16, Pl. 7, Figs. 9, 10.
?*part.* 1975 *Thamnophyllum truktense* (sic) (McLaren); Besprozvannykh et al., p. 57, Pl. 12, Figs. 2, 3.

non 1977 *Thamnophyllum tructense* (McLaren); Spasskiy, Pl. 12, Figs. 2a, b.

Type material. GSC 6310 (holotype); Kakisa Formation, 24.6–25.9 m below top, immediately below Whittaker Falls, Trout River, District of Mackenzie, locality 3 (Loc. 15 of Smith 1945, p. 71) of Locality Register and Fig. 4. GSC 6309 ("hypotype" — McLaren 1959, p. 30); upper Kakisa Formation, Coral Falls, Trout River, District of Mackenzie, locality 74 (Loc. 20 of Smith 1945, p. 71) of Locality Register and Fig. 4. The other District of Mackenzie specimens included in "*D.* (*S.*) *stramineum*" by Smith (1945), GSC 9337, 9338 and 9341 (? = 6203), are here included in *Peneckiella metalinae* (see below).

Additional material. GSC 126394; Kakisa Formation, 44.8 m below top, near lowest falls in Trout River gorge, District of Mackenzie, locality 3 (Amoco Loc. 11163) of Locality Register and Fig. 4. GSC 126395, 126396 (topotypes); Kakisa Formation, 24.6–25.9 m below top, immediately below Whittaker Falls, Trout River, District of Mackenzie, locality 3 (Amoco Loc. 11163) of Locality Register and Fig. 4. GSC 126397; Kakisa Formation, 13–18 m above base, 20–25 m below top, "Ottertail Ridge", British Columbia, locality 115 (Amoco Loc. 14055) of Locality Register and Fig. 7. GSC 126398; Simla Formation, 40.4–65.4 m above base, 0.0–25.0 m below top, Persimmon Creek, Alberta, locality 116 (Amoco Loc. 15252) of Locality Register and Fig. 9. GSC 126399; Simla Formation, 69.8–76.5 m above base, 0.0–6.7 m below top, northwest flank of Mt. Mackenzie, Alberta, locality 28 (Amoco Loc. 11496) of Locality Register and Fig. 10.

Population studied. 16 sectioned coralla.

Range. Latest Frasnian, zone 13, *P. variabilis – S. cinctus* coral faunas.

Diagnosis. Corallum phaceloid, with periodic outgrowths of dissepimentarium linking adjoining corallites, so that corallum may be locally subcerioid to cateniform. Corallite diameter 3.2–4.0 mm, with major septa ranging from 13 to 16 in number. Both major and minor septa very short, major septa only slightly longer than minor, barely extending into tabularium. Horseshoe dissepiments very small, inner dissepiments absent, outer dissepiments very variably developed. Where present, outer dissepiments are flat, concave, or globose to elongate, particularly where outgrowths of dissepimentarium connect adjoining corallites. Tabulae mainly complete, flat or variably sagging.

Description. Corallum is mainly phaceloid, with corallites generally closely spaced. Periodically, the outer dissepimentarium expands laterally so that adjoining corallites are in contact, and corallum becomes subcerioid to cateniform. Corallite diameter ranges from 3.2 to 4.0 mm in available material, with a mean of 3.5 mm. Major septa vary from 13 to 16 in number (mean 14). Both major and minor septa show moderate to strong fusiform dilation, with dilation slightly more pronounced in major septa. Coarse rhipidacanthine trabeculae are typically present, with spinose tips in places appearing as dots in longitudinal sections of tabularium adjacent to horseshoe pipe. Major septa are very short, typically barely extending into tabularium, rarely slightly longer. Minor septa are only slightly shorter and are confined to dissepimentarium.

Horseshoe dissepiments are small, mainly form a continuous pipe, and are variably coated with stereome. In some places the horseshoes may be partly unrolled, and become peneckielloid or globose. However, this is very rare and never to the extent commonly seen in the genus *Peneckiella* (see discussion of that genus, below). Inner dissepiments do not appear to be present, but outer dissepiments may be developed. If the latter are preserved, they are flat, concave, or, most commonly, very small and globose to elongate. Where the outer dissepimentarium expands to connect adjoining corallites, dissepiments are typically globose to elongate. Tabulae are mainly complete, generally flat and fairly closely spaced (usually 3–4/mm). In more rare cases, tabulae may form sagging series, although this may vary even in the one corallite (e.g., GSC 126396, Pl. 11, Fig. 4).

Remarks. This distinctive species was first described by Smith (1945, pp. 23–24) under the name *Disphyllum* (*Synaptophyllum*) *stramineum* (Billings, 1859). Revision of *Diphyphyllum stramineum* Billings, 1859, from the Eifelian Edgecliff Member, Onondaga Formation of southern Ontario by McLaren (1959) and Oliver (1976) has led to its present assignment to the genus *Acinophyllum* McLaren, 1959, a craspedophyllid unrelated to the phillipsastreid *Thamnophyllum*. As discussed by McLaren (1959, pp. 29–30), Smith's material of "*D.* (*S.*) *stramineum*" includes more than one species. McLaren proposed the name *Phacellophyllum tructense* for the specimens Smith regarded as "typical" or "usual", which are GSC 6310 (holotype) and GSC 6309, from the late Frasnian Kakisa Formation on Trout River, southern District of Mackenzie. *T. tructense* is very common at that locality and often forms small patch reefs, as illustrated by McLean et al. (1987, Pl. 9; 1995, Pl. 9). The other material assigned to "*D.* (*S.*) *stramineum*" by Smith is from other horizons and localities, and is here placed in *Peneckiella metalinae* Sorauf, 1972 (see discussion of that species, below).

T. tructense is an unusual form in having common extension of the outer dissepimentarium so that adjoining corallites may be variably in lateral contact. In this feature the species shows similarities to the genus *Smithicyathus* Różkowska, 1980, especially forms of that primarily cerioid to subcerioid genus in which corallites become separated and partially phaceloid. This character is shown in some specimens of *S. cinctus* (Smith 1945), revised by McLean (1994*b*), and also present in the Kakisa Formation of southern District of Mackenzie. While having considerably larger corallites than *T. tructense*, and thus not likely to be confused with that species, *S. cinctus* may show general morphological similarities to specimens of *T. tructe*nse which have extreme development of connecting outer dissepiments (compare McLean 1994*b*, Pl. 2, Figs. 3, 5 with Pl. 10, Figs. 2, 8 and Pl. 11, Fig. 1, herein). While *T. tructense* could be considered as a form transitional between typically cerioid to subcerioid *Smithicyathus* and typically phaceloid *Thamnophyllum*, it is placed here in *Thamnophyllum* since the phaceloid growth form is much more dominant than the subcerioid.

Soshkina (1952, p. 85) mentioned "*T. stramineum* (Billings)" as occurring in the Frasnian of Armenia and Canada, presumably meaning that she had Armenian material she felt was conspecific with Smith's (1945) concept of *stramineum*, and thus potentially representative of *T. tructense*. However, as her material apparently had 18–20 major septa, as compared to 13–16 in *T. tructense*, the Armenian form, unfortunately not illustrated, is probably separable from *T. tructense*. Similarly, Spasskiy (1960, p. 48, Pl. 15, Figs. 1–3) assigned a form from the Frasnian Kamyshinkin Beds of the Rudniy Altai to "*T. stramineum*" (later referred to as *T. tructense* by Spasskiy 1977). This material can be clearly separated from *T. tructense* by having longer major septa and thus more complex tabulae, as well as not appearing to have the complex outer dissepimentarium of the Canadian species.

Material assigned to *T. tructense* (misspelled as *truktense*) by Besprozvannykh et al. (1975, p. 57, Pl. 12, Figs. 2, 3), from the Frasnian of Sette-Daban and the New Siberian Islands, northeastern Russia, in part shows close similarities to the Canadian species. The specimen they illustrated in Pl. 12, Figs. 2a, b, in particular, is comparable in corallite size, septal number and length, and character of the tabularium. The other specimen (Pl. 12, Fig. 3), illustrated only in transverse section of a single corallite, has considerably longer septa than *T. tructense*, and Besprozvannykh et al. (1975, p. 57) specifically stated that the major septa in their material usually extend up to 2/3 of the corallite radius. Additionally, presence of abundant outer dissepiments, although very variably developed in the Canadian specimens, is neither mentioned nor illustrated in the Russian material. While it is possible more than one species (short and long septate) may be present in the Russian material, only the short septate form could be considered as a possible representative of *T. tructense*. Further study of the Russian fauna is clearly needed, with assessment of its variability.

Of other Canadian representatives of *Thamnophyllum*, *T. pedderi* n. sp. has a similar corallite size to *T. tructense*. However, although the major septa of *T. pedderi* are mostly relatively short, they show greater variation in length than in *T. tructense*. Additionally, the tabulae of *T. pedderi* are more widely spaced and there does not appear to be a tendency to develop abundant outer dissepiments.

Thamnophyllum pedderi n. sp.

Pl. 11, Figs. 5–9; Pl. 12, Figs. 1–8; Pl. 13, Figs. 1–6

1956 *Synaptophyllum* cf. *stramineum* (Billings); McLaren, p. 53.

Derivation of name. In honour of A.E.H. Pedder, for his contribution to the study of Canadian Devonian rugose corals.

Type material. GSC 126400 (holotype); upper Nisku Formation at 8317.2 ft (2535.0 m), Amoco Chevron B-1 Iosegun well, Lsd 7, Sec. 32, Twp 65, Rge 19 W5M, Alberta (Well loc. 3 of Fig. 15). GSC 126401 (paratype); Ireton Formation at 12343 ft (3762.0 m), BA Shell Berland River well, Lsd 10, Sec. 32, Twp 58, Rge 23 W5M, Alberta (Well loc. 11 of Fig. 15). GSC 126402 (paratype); upper Nisku Formation at 7371.5 ft (2246.7 m), Huber NCO Mesa Allan well, Lsd 10, Sec. 3, Twp 68, Rge 14 W5M, Alberta (Well loc. 5 of Fig. 15). GSC 126403 (paratype); upper Nisku Formation at 5282 ft (1609.9 m), Horn River et al. Majeau Lake well, Lsd 4, Sec. 29, Twp 57, Rge 3 W5M, Alberta (Well loc. 6 of Fig. 15). GSC 126404 (paratype); Bigoray Member, Nisku Formation at 2353.7 m, Amoco et al. Bigoray well, Lsd 16, Sec. 4, Twp 52, Rge 8 W5M, Alberta, (Well loc. 15 of Fig. 15). GSC 126405 (paratype); Mt. Hawk Formation, 113.4–115.5 m above base, 8.8–11.0 m below top, Roche Miette, Alberta, locality 117 (GSC Loc. 36863) of Locality Register and Fig. 10. GSC 16722, 16723 (paratypes); Cripple Tongue, 18.0–30.5 m above base, 7.9–20.4 m below top, Job Creek headwaters, Alberta, locality 121 (GSC Loc. 25176) of Locality Register and Fig. 11. GSC 126406 (paratype); Cripple Tongue, 33.8 m above base, 52.1 m below top, Allstones Creek headwaters, Alberta, locality 66 (Amoco Loc. 11522) of Locality Register and Fig. 12. GSC 126407 (paratype); upper member, Cairn Formation, basal 32.3 m, ridge 4 km west of North Burnt Timber Creek, Alberta, locality 66 (Amoco Loc. 11522) of Locality Register and Fig. 12. GSC 126408 (paratype); Cairn Formation, 60.7 m below top, Mt. McDougall, Alberta, locality 67 (Amoco Loc. 11520) of Locality Register and Fig. 13.

Population studied. 74 sectioned coralla.

Range. Mid – late Frasnian, zones 6–12, *A. parvulum* – *H. magna* coral faunas.

Diagnosis. Dendroid to phaceloid *Thamnophyllum* with corallite diameter 2.5–5.3 mm and 12–18 major septa. Major septa are highly variable in length, although generally short, mainly 0.2–0.6 of corallite radius. Peripheral septal dilation is also highly variable, ranging from very slight to strong. Horseshoe dissepiments are small, in regular pipe. Inner dissepiments are absent, outer dissepiments only developed adjacent to lateral offsets. Tabulae are dominantly complete and flat, with highly variable spacing.

Description. Corallum is dendroid to phaceloid and typically fragmented in most colonies. Adult corallite diameter ranges from 2.5 to 5.3 mm, with most common value of 3.0–4.5 mm and mean of 3.7 mm. Major septa vary in number from 12 to 18, with 14–16 most typical, and a mean of 16. Length of major septa is extremely variable, generally ranging from about 0.2 of corallite radius (barely extending into tabularium) to 0.6 of radius. In a few cases they may extend up to 0.8 of radius and commonly the length will vary greatly within the one corallum, and even the one corallite. A range in length of about 0.3–0.5 of radius is most common in the studied population. Minor septa are very short and primarily confined to the peripheral pipe of horseshoe dissepiments. Both orders of septa exhibit great variation in peripheral fusiform dilation, some specimens showing very little dilation or stereome coating of horseshoes and septa (e.g., GSC 126400, Pl. 12, Figs. 1–3), while others have quite pronounced dilation (although unevenly developed) and peripheral stereome coating (e.g., GSC 126404, Pl. 12, Figs. 4–6). All variations between these two extremes are seen within the population. Rhipidacanthine trabeculae are well developed within the area of the horseshoe pipe, most clearly seen in specimens with more dilated septa.

Horseshoe dissepiments are typically small and form a continuous pipe. Inner dissepiments are absent, as are outer dissepiments generally, except in some cases at corallite lateral offsets. In the latter situations, a few small, globose, outer dissepiments may occur joining the lower part of the offset to the parent corallite (e.g., GSC 126404, Pl. 12, Fig. 5). Tabulae are mostly complete and commonly flat; in rarer cases they may be slightly arched or sagging. Tabular spacing is highly variable: it may range from 0.3 to 2.0 mm, but values in the range of 0.8–1.2 mm are most common.

Remarks. T. pedderi n. sp. shows general similarities to "*Peneckiella*" *achanayensis* Soshkina, 1939, from the Frasnian of the Akhanay River, western slopes of the southern Urals. Unfortunately, this species has been illustrated only by line drawings of the holotype (Soshkina 1939, Pl. 9, Figs. 74, 75), but these show the variable, although generally short, major septa, single row of small dissepiments, and mainly simple tabulae of the Canadian form. In addition, the corallite diameter and septal number are within the range of *T. pedderi*. However, uncertainty as to the nature of the dissepiments in "*P*". *achanayensis* precludes assigning the Canadian material to that species. Some of the dissepiments in Soshkina's drawing (Pl. 9, fig, 75) do appear to be horseshoe-shaped, but others appear more globose, although apparently trabeculae or stereome partly obscure them. According to Soshkina (1939, p. 51), the dissepiments are "normal", presumably meaning globose, and as she recognized horseshoes in other species (e.g., "*Peneckiella*" *monozonata*), she apparently did not consider them to be present in "*P*". *achanayensis*. Scrutton (1968, p. 274) regarded the dissepiments to be peneckielloid, and consequently retained the species in *Peneckiella*. They do not appear peneckielloid, as that term is interpreted by the present writer (see discussion of *Peneckiella*, below), based solely on Soshkina's drawing, but a variety of styles could be present and restudy of the type (and preferably additional) material is necessary to determine this. Ivanovskiy and Shurygina (1980) included "*P*". *achanayensis* in *Thamnophyllum monozonatum* (Soshkina 1939), also from the Frasnian of the western slopes of the southern Urals, and, as discussed above, regarded here as a synonym of *Thamnophyllum colemanense* (Warren, 1928). *T. monozonatum* has well-developed horseshoe dissepiments, noted by Ivanovskiy and Shurygina (1980, p. 23) in their revision of the species, and clear in Soshkina's drawing of the holotype (1939, Pl. 10, Fig. 77). The inference from this assignment is that Ivanovskiy and Shurygina believed that "*P*". *achanayensis* possesses horseshoe dissepiments. With such uncertainty as to the nature of the dissepiments of "*P*". *achanayensis*, confident generic assignment of the species is not possible at present. If "*P*". *achanayensis* should prove to be a representative of *Thamnophyllum* (i.e., by demonstration that it has a continuous pipe of horseshoe dissepiments), then it would need to be considered as a possible senior synonym of *T. pedderi*. However, for the present it is felt to be preferable to assign the Canadian material to a separate species, whose variability can be well characterized by the large population available for study.

Distinction of *T. pedderi* from the most similar Canadian species, *T. colemanense* (Warren) and *T. tructense* (McLaren), is discussed above under remarks on those species. The combination of relatively small corallites, short major septa, absence of inner, and, for the most part, outer dissepiments, together with simple, flat tabulae, serves to distinguish *T. pedderi* from all other described species.

Thamnophyllum cordense n. sp.
Pl. 13, Figs. 7, 8; Pl. 14, Figs. 2, 4

Derivation of name. Latin, cor, cordis = heart, a reference to Heart Lake, District of Mackenzie, adjacent to the type locality.

Type material. GSC 126409 (holotype); Alexandra Member, Twin Falls Formation, roadcut 3 m thick, Mackenzie Highway, District of Mackenzie, locality 120 (uncatalogued) of Locality Register and Fig. 5.

Population studied. 1 corallum.

Range. Mid Frasnian, ?zone 10 – early zone 11, *Temnophyllum* sp. A coral fauna.

Diagnosis. Thamnophyllum with large corallites, diameter ranging from 14 to 18 mm, and 29–35 major septa. Major septa show significant peripheral dilation, becoming thin and variably contorted axially; they extend 0.8–0.9 of corallite radius. Horseshoe dissepiments form a continuous pipe, are very small, and generally coated with stereome. Numerous small, globose dissepiments typically line both sides of the horseshoe pipe. Occasional flat or concave outer dissepiments may also occur. Tabulae are typically complete and form axially sagging series.

Description. Corallum is fasciculate with peripheral increase. Corallites are generally closely spaced and occasionally have a slightly elliptical transverse section. Corallite diameter ranges from 14 to 15 mm in circular corallites, reaching 18 mm in more elliptical forms. Major septa range in number from 29 to 35 in mature corallites, with a mean of 32 from 9 corallites of the single corallum available for study. They show moderate fusiform dilation in the dissepimentarium and outer tabularium, becoming very slender and variably contorted in the inner tabularium. Overall, they range in length 0.8–0.9 of corallite radius. Minor septa show less pronounced fusiform dilation and range up to about one third the length of the major septa. Well-preserved rhipidacanthine trabeculae can be recognized in peripheral, dilated parts of septa.

Horseshoe dissepiments are typically small and form a continuous pipe; they commonly have a thin coating of stereome. One to four rows of small, globose dissepiments typically line the inner side of the horseshoe pipe and similar rows usually line the outer side. In some corallites the outer dissepiments may be larger, more elongate, and, in some cases, flat to sagging. Tabulae are incomplete peripherally and are in flat to slightly depressed series. Axially they are more complete and mainly in sagging series.

Remarks. Unfortunately only one corallum of this species was available for study, but it is well preserved and so clearly distinctive from other described species, that erection of a new species appears justified. Of Canadian forms, only *T. julli* n. sp. has similarities to the new species. These can be seen in the large corallite diameter, septal number, and axial contortion of the major septa. Clear distinction can be made, however, by *T. cordense* having more closely spaced corallites, greater peripheral septal dilation, smaller horseshoe dissepiments, and abundant inner and outer dissepiments.

The most closely related species appears to be *T. supradevonicum* Penecke, 1904, from the Frasnian of the Antitaurus Mountains, Turkey. This species was proposed as type for the new genus *Pseudopexiphyllum* by Hubmann (1992), but reasons for retaining it in *Thamnophyllum* are given in the generic discussion of *Thamnophyllum*, above. From the work of Penecke (1904, pp. 144–146; Pl. 4, Figs. 1a, b) and Hubmann (1992, pp. 161–162; Pl. 2, Figs. 3–6; Text-Fig. 8a, b), similarities to *T. cordense* may be seen in corallite diameter, tabular shape, and the presence of globose inner and outer dissepiments. However, *T. supradevonicum* can be clearly distinguished by having fewer and less dilated septa, larger horseshoe dissepiments, and a much wider zone of outer dissepiments, which appear to be globose only and not the mix of globose and planar forms of *T. cordense.*

Thamnophyllum julli n. sp.
Pl. 14, Figs. 1, 3, 5–8; Pl. 15, Figs. 1–6, 8, 9

1998 *Thamnophyllum* sp. A. McLean and Klapper, Pl. 1, Figs. 2–4.

Derivation of name. In honour of the late R.K. Jull, coral worker and collector of much of the outcrop material of this species.

Type material. GSC 126410 (holotype); Zeta Lake Member, Nisku Formation at 2390.8 m, Amoco et al. Bigoray well, Lsd 13, Sec. 9, Twp 52, Rge 8 W5M, Alberta (Well loc. 13 of Fig. 15). GSC 126411 (paratype); Zeta Lake Member, Nisku Formation at 2393.1 m, same well as holotype, GSC 126410, above. GSC 126412 (paratype); Zeta Lake Member, Nisku Formation at 2350.2 m, Amoco et al. Bigoray well, Lsd 1, Sec. 7, Twp 52, Rge 8 W5M, Alberta (Well loc. 14 of Fig. 15). GSC 126413, 126414 (paratypes); upper Southesk Formation, 2.4 m thick unit of thickly bedded limestone, approximately 12.2 m below base of Simla Formation, west side of Mt. Haultain cirque, Alberta, locality 27 (R.K. Jull Loc. L-199) of Locality Register and Fig. 9. GSC 126415 (paratype); undifferentiated Southesk Formation, upper Mt. Hawk Formation equivalent, 96.9–106.1 m below base of Palliser Formation, ridge northwest of Mt. Gregg, Alberta, locality 59 (Amoco Loc. 11423) of Locality Register and Fig. 10.

Population studied. 23 sectioned coralla.

Range. Late Frasnian, zone 12, *M. pustulosa – H. magna* coral faunas.

Diagnosis. Thamnophyllum characterized by large corallites, common diameter 12–14 mm, and with generally 25–29 major septa. Septa are typically thin, both axially and peripherally, with common stereome coating in peripheral areas. Major septa range from 0.6 to 0.9 of corallite radius, and are typically contorted axially. Horseshoe dissepiments are large and form a continuous pipe. Outer dissepiments are rarely developed; if present they are flat or small and globose. Tabulae are widely spaced, in flat to variably arched series.

Description. All coralla available for study are fragmentary, but clearly fasciculate, with peripheral increase and generally well-spaced corallites. Corallite diameter ranges from 9 to 18 mm in adult stages, with most common values of 12–14 mm and a mean of 12.7 mm. Major septa vary in number from 22 to 32 in adult specimens, with 25–29 being most common (mean 27). Both major and minor septa are thin, with only slight peripheral fusiform dilation. Peripheral ends of septa are typically coated with stereome. Major septa extend from 0.6 to 0.9 of corallite radius and commonly show variable degrees of contortion in axial region. Minor septa are generally 0.2–0.4 of length of major septa. Rhipidacanthine trabeculae are clearly recognizable in peripheral, more dilated parts of septa.

Horseshoe dissepiments are large and form a continuous pipe; typically they are coated with stereome. Inner dissepiments are not commonly developed; if present, they are large and in a single, discontinuous row. Outer dissepiments appear to be generally not developed; only rare examples are seen and are both flat and small, globose forms lining the outside of the horseshoe pipe. Tabulae are mostly incomplete and widely spaced, generally 1–2/mm. They form mainly flat series in shorter septate corallites, but sag peripherally and have broad axial arches in longer septate forms, in some cases with an axial sag as well.

Remarks. The large corallite dimensions and septal number of *T. julli* serve to distinguish it from most other representatives of *Thamnophyllum*. The only other large Canadian species, *T. cordense* n. sp., is clearly separated by having greater peripheral septal dilation, much smaller horseshoe dissepiments, and abundant globose inner and outer dissepiments. *T. hollardi* Coen-Aubert, 1982, from the red marble reef facies "F2j", late Frasnian, Belgium, has some axial contortion of the major septa and large horseshoes as in *T. julli*. However, it may be distinguished from *T. julli* by slightly smaller dimensions (corallite diameter 7–11 mm, 18–23 major septa), greater peripheral septal dilation, and more sagging series of tabulae (Coen-Aubert 1982, pp. 35–37, Pl. 6, Figs. 39–41).

Genus *Peneckiella* Soshkina, 1939

Peneckiella Soshkina, 1939, p. 23.
Sudetia Różkowska, 1960, p. 35.

Type species. Diphyphyllum minus Roemer, 1855.

Diagnosis. Corallum fasciculate, with lateral and possibly axial increase. Septa have varying degrees of peripheral dilation, with rhipidacanthine trabeculae in full and partial fans. Dissepiments in one to four rows, rarely more, peneckielloid and globose, with occasional horseshoes. Tabulae complete and incomplete, mainly flat to strongly arched.

Discussion. Soshkina (1939) gave a very broad interpretation to the genus when erecting *Peneckiella*, including solitary, branching, and massive forms. Most subsequent authors have restricted it to branching forms only, as exemplified by the type species, *P. minor* (Roemer 1855), from the mid Frasnian Ibergerkalk at Winterberg, near Bad Grund, Harz Mountains, Germany, and this view is followed here. The age of the type stratum is inferred from data given by Gischler (*in* Gischler et al. 1991, p. 25), who referred to abundant representatives of *P. minor* in beds in the Winterberg Quarry, stated to be of mid Frasnian "lower *hassi* zone" age (equivalent to Montagne Noire zones 7–8, Klapper and Becker, 1999). The character of the dissepiments in *P. minor* is fundamental to the understanding of the genus, together with the nature of the trabeculae. At the time of proposal of *Peneckiella*, the internal morphology of *P. minor* was known only from the drawings of Roemer (1855, Pl. 6, Figs. 12a–c), Frech (1885, Pl. 1, Figs. 3, 3a, 3b) and Lang and Smith (1935, Pl. 35, Fig. 3). These all show a single row of dissepiments, which are mainly globose, but which also have a tendency for their upper surface to slope downward at the contact with the corallite wall. This style, clearly illustrated diagrammatically by Strusz (1965, Text-Fig. 15b), is known as peneckielloid. Strusz (1965, Text-Fig. 15c) also illustrated forms referred to as sigmoidal, but these are regarded here as variants of the peneckielloid type, and are not separated from that group. While peneckielloid dissepiments are present in the early illustrations of *P. minor*, they were not recognized as significant by Soshkina (1939) in diagnosing *Peneckiella*. Subsequent authors have studied topotypes of *P. minor* and noted the variability of dissepiment styles. Flügel (1956) regarded the dissepiments to be only horseshoes, Schouppé (1958, Figs. 12–14) illustrated globose and horseshoe forms, while Pickett (1967, Fig. 14) showed a mixture of horseshoe and peneckielloid dissepiments. Scrutton (1968, Pl. 17, Figs. 4, 5) provided photographic illustration of thin sections of the holotype for the first time, but while a few globose dissepiments are visible, the specimen is too poorly preserved for the variation to be clearly seen. Fortunately the species is quite common at the type locality, as mentioned above, and the present writer collected several specimens there in 1991. The material is well preserved and sections are illustrated here (Pl. 15, Fig. 7; Pl. 16, Figs. 1–6; Pl. 17, Fig. 1) to show development of peneckielloid, globose, and horseshoe dissepiments, in addition to rhipidacanthine trabecular fans, which previously have not been shown photographically.

The genus *Sudetia* Różkowska, 1960, based on *S. lateseptata* Różkowska, 1960, from the late Frasnian of Mokrzeszów, Sudeten Mountains, Poland, has globose and peneckielloid dissepiments, together with well-developed trabecular fans (Różkowska 1960, Figs. 31–33). The present writer agrees with Scrutton (1968) in regarding *Sudetia* as a junior synonym of *Peneckiella*. Scrutton (1968) also regarded *Acinophyllum* McLaren, 1959 as a synonym of *Peneckiella*. However, revision of the type species of *Acinophyllum*, *A. simcoense* (Billings 1859), from the late Emsian Bois Blanc Formation of Ontario, by Oliver (1976) in a thorough study of that genus, has shown that *Acinophyllum* has only globose dissepiments, with fine, monocanthine trabeculae. As noted by Oliver, *Acinophyllum* is a craspedophyllid, unrelated to the phillipsastreid *Peneckiella*.

Zelolasma Pedder, 1964 (type species *Z. gemmiforme* (Etheridge 1902), Emsian, Cavan Limestone, New South Wales) has globose, peneckielloid, and horseshoe dissepiments (Pedder 1964, Pl. 19, Fig. 2; Pickett 1967, Pl. 7, Fig. 27), but can be distinguished from *Peneckiella* by having relatively fine trabeculae, which can give "ragged, locally acanthine adaxial septal edges" (Pedder 1998, p. 237). *Asarcophyllum* Pedder, 1998 (type species *A. ramosum* Pedder, 1998, Emsian, Bartine Member, McColley Canyon Formation, Nevada) is also similar to *Peneckiella* in developing globose, peneckielloid, and rare horseshoe dissepiments. It may be separated by having very slender septa, which are non-trabeculate (Pedder 1998, Figs. 7.1–7.8, 8.1–8.4).

A final genus to be considered here is *Parapeneckiella* Wang, 1994, based on *P. shidiaensis* Wang, 1994, from the Heyunzhai Formation (Frasnian) of Yunnan. It was considered by Wang to differ from *Peneckiella* by having solitary coralla, with some amplexoid septa. This is certainly true, but *P. shidiaensis* shows typical morphology (Wang 1994, Pl. 4, Fig. 3) of the solitary charactophyllid *Chostophyllum* Pedder, 1982, and *Parapeneckiella* is probably a synonym of that genus.

Two Pragian species from New South Wales referred to *Peneckiella* by Strusz (1965) and other authors raise some questions as to their correct generic assignment. *Disphyllum mesa* Hill, 1942*a* and *Peneckiella boreensis* Strusz, 1965, both occurring in the Garra Formation of the Wellington area, have everted dissepimentaria of globose, peneckielloid, and horseshoe dissepiments, and rather fine trabecular fans are present in "*D*". *mesa* (Strusz 1965, Text-Fig. 16b; Pl. 77, Fig. 3; Pl. 78, Figs. 1, 2). However, both these species have axial, parricidal increase, while the Frasnian species referred here to *Peneckiella* (see list below) appear to have only peripheral, non-parricidal increase. Combining this feature with the time gap between these two Pragian species and the undoubted representatives of *Peneckiella*, the oldest of which appear to be Givetian (*P. pyrenaica* Joseph and Tsien, 1975 and its possible synonym *P. ramondi* Joseph and Tsien, 1975, from the Pyrénées of southwest France), the question arises as to whether the Pragian species may be early homeomorphs of *Peneckiella*, or possibly related to *Zelolasma* (which also has parricidal, axial increase). Restudy of *Z. gemmiforme*, "*D*". *mesa*, and "*P*". *boreensis*, especially with regard to their septal microstructure, appears necessary to resolve this issue.

Soshkina (1939) erected a number of new Frasnian species from the western slopes of the southern Urals, which she assigned to *Peneckiella*. In the present writer's opinion, none can be regarded confidently as congeneric with *P. minor*. As noted above, the nature of the dissepiments in *P. achanayensis* Soshkina, 1939 is uncertain, and it could belong to *Thamnophyllum* or *Peneckiella*. *P. monozonata* Soshkina, 1939 has a well-developed pipe of horseshoe dissepiments and is a clear representative of *Thamnophyllum*; it is considered here to be synonymous with *T. colemanense* (Warren 1928), as discussed above. *P. spiralis* Soshkina, 1939 also appears to have continuous horseshoes, but it is uncertain if it is solitary or branching (see discussion of *T. colemanense*, above). *P. nalivkini* Soshkina, 1939 does appear to have a tendency to develop peneckielloid dissepiments, but the line drawing of a longitudinal section of a single corallite (Soshkina 1939, Pl. 8, Fig. 69) does not show this clearly enough to place this species in *Peneckiella* without reservation. It requires further study and is regarded here as only a possible representative of the genus.

A number of other species were originally assigned to *Peneckiella*, but appear to be clearly separable. As noted above, Soshkina (1939) included cerioid forms in her original concept of *Peneckiella*, and various species with that growth form have been referred to the genus. In this category, and excluded from *Peneckiella*, are *P. darwini* (Frech 1885) *sensu* Soshkina (1939) (Frasnian, western slopes of southern Urals, ? = *Columnaria* Goldfuss, 1826), *P. juresanensis* Soshkina, 1951 (Givetian, western slopes of southern Urals, revised by Ivanovskiy and Shurygina 1980 and assigned to *Trapezophyllum* Etheridge, 1899), *P. tabulata* Bulvanker *in* Ivaniya 1952 (Glubok Beds, late Frasnian, Kuznets Basin), *P. porfirjevi* Spasskiy, 1955 (Vyazov Beds, Emsian, southern Urals), *P. baschkirica* Spasskiy, 1955 (Biya Beds, ?late Emsian – early Eifelian, southern Urals) and *P. primitiva* Sayutina, 1965 (Givetian, Armenia). The subcerioid *P. ?intermedia* Cao *in* Cao and Ouyang 1987, from the Frasnian Cakuohe Formation of the Qinling Mountains, China, is likewise excluded from *Peneckiella* on growth form alone. It may be related to *Thryptophyllum* Pedder, 1990.

Some branching forms should also be excluded from *Peneckiella*. They all appear to lack peneckielloid dissepiments, having axially sloping rows of globose forms, and are more likely to be disphyllids. Included in this group are *P. tolstichinae* Soshkina, 1952 (Voronezh Beds, Frasnian, Russian Platform), *P. jevlanensis* Bulvanker *in* Soshkina 1952 (Liven Beds, Frasnian, Russian Platform), *P. irregularis* Ivaniya, 1965 (Glubok Beds, late Frasnian, Kuznets Basin), *P. zhongguoensis* Jia, 1977 (Sipai Formation, Emsian, Guangxi) and *P. raritabulata* Jia, 1977 (Middle Devonian, Guangxi).

The following species are regarded here as representative of *Peneckiella*. After the type species they are listed in chronological order of publication.

Diphyphyllum minus Roemer, 1855. Type species. Ibergerkalk, mid Frasnian, Winterberg, Harz Mountains, Germany. Holotype illustrated by Scrutton (1968). Topotypic material illustrated by Schouppé (1958), Pickett (1967), and herein.

Cyathophyllum kunthi Dames, 1869 *sensu* Różkowska (1957) and Fedorowski (2003). Late Frasnian, Sudeten Mountains, Poland. Not *C. kunthi* Dames *sensu* Dybowski (1873) = *Thamnophyllum*.

Cylindrophyllum floydense Belanski, 1928. Mason City Member, Shellrock Formation, mid Frasnian, Iowa. Revised by Sorauf (1998) and herein.

Spinophyllum fasciculare Soshkina, 1939. Early Frasnian, western slopes of southern Urals. Regarded here as a synonym of *P. floydensis*.

Disphyllum (*Synaptophyllum*) *densum* Smith, 1945. Upper Member, Twin Falls Formation, mid Frasnian, southern District of Mackenzie. Revised herein and regarded as a synonym of *P. floydensis*.

Peneckiella teicherti Hill, 1954. Sadler Limestone, Frasnian, Canning Basin, Western Australia. Holotype reillustrated by Strusz (1965) and species revised by Hill and Jell (1970).

Schlueteria lyskovensis Ermakova, 1957. Upper Voronezh Horizon, late Frasnian, Russian Platform.

Sudetia lateseptata Różkowska, 1960. Late Frasnian, Sudeten Mountains, Poland. Type species of *Sudetia* Różkowska, 1960.

Disphyllum caespitosum tryciclicum Schouppé, 1965. Frasnian, Chitral, Pakistan.

Peneckiella salternensis Scrutton, 1968. Frasnian, Devon.

Peneckiella metalinae Sorauf, 1972. Mid Frasnian, Washington. Revised below.

Peneckiella pyrenaica Joseph and Tsien, 1975. Early Givetian, western Pyrénées, southern France.

Peneckiella ramondi Joseph and Tsien, 1975. Early Givetian, western Pyrénées, southern France. Possibly synonymous with *P. pyrenaica*.

Peneckiella szulczewskii Różkowska, 1980. Early Frasnian, Holy Cross Mountains, Poland. Regarded here as a synonym of *P. floydensis*.

Peneckiella sachaninensis Lakhov, 1986. Zhandr Horizon, early Frasnian, Novaya Zemlya.

Peneckiella guangxiensis Yu and Kuang, 1986. Rongxian Formation, late Frasnian, Guangxi. Regarded here as a synonym of *P. metalinae*.

Peneckiella liujingensis Yu and Kuang, 1986. Rongxian Formation, late Frasnian, Guangxi.

Peneckiella yunnanensis Wang, 1994. Heyuanzhai Formation, Frasnian, Yunnan.

Peneckiella duponti Coen-Aubert, 1995. Lion Member, Grands Breux Formation, mid Frasnian, Dinant Basin, Belgium. Regarded here as a synonym of *P. metalinae*.

Peneckiella gracilis n. sp. Cripple Tongue, mid Frasnian, west central Alberta Rocky Mountains.

Peneckiella haultainensis n. sp. Simla Formation, late Frasnian, west central Alberta Rocky Mountains.

Species listed below may belong to *Peneckiella*, but require further study to confirm such an assignment.

Peneckiella nalivkini Soshkina, 1939. Late Frasnian, western slopes of southern Urals.

Peneckiella achanayensis Soshkina, 1939. Late Frasnian, western slopes of southern Urals. See also remarks above on *Thamnophyllum pedderi*.

Disphyllum mesa Hill, 1942*a*. Garra Formation, Pragian, Wellington area, New South Wales. Revised by Strusz (1965).

Macgeea langi Sun, 1958. Shetianqiao Formation, Frasnian, Hunan. See discussion of *Macgeea*, above.

Hexagonaria yakovlevi Bulvanker, 1958. Glubok Beds, late Frasnian, Kuznets Basin.

Hexagonaria isylica Bulvanker, 1958. Glubok Beds, late Frasnian, Kuznets Basin.

Peneckiella belskayae Ivaniya *in* Zheltonogova and Ivaniya 1961. Glubok Beds, late Frasnian, Kuznets Basin.
Peneckiella glubokiensis Ivaniya *in* Zheltonogova and Ivaniya 1961. Glubok Beds, late Frasnian, Kuznets Basin.
Peneckiella carinata Ivaniya, 1965. Solomino Beds, late Frasnian, Kuznets Basin.
Peneckiella boreensis Strusz, 1965. Garra Formation, Pragian, Wellington area, New South Wales.
Peneckiella regularis Jia, 1977. Shetianqiao Formation, Frasnian, Hunan.
Petrozium zhongguoense Jia, 1977. Frasnian, Hubei.
Peneckiella shawoziensis He, 1978. Shawozi Formation, Frasnian, Sichuan.
Peneckiella longiseptata He, 1978. Shawozi Formation, Frasnian, Sichuan.
Peneckiella melanopotama Lakhov, 1981. Late Frasnian, Novaya Zemlya.
Peneckiella xizangensis Liao *in* Wu et al. 1982. Frasnian, Xizang.
"*Peneckiella*" *raritabulata* Fan *in* He and Fan 1988. Tuqiaozi Formation, Frasnian, Sichuan.
"*Peneckiella*" *irregularis* Fan *in* He and Fan 1988. Shawozi Formation, Frasnian, Sichuan.

Peneckiella floydensis (Belanski, 1928)

Pl. 17, Figs. 2, 5; Pl. 18, Figs. 1–6; Pl. 19, Figs. 1–7; Pl. 20, Figs. 1–4

1928 *Cylindrophyllum floydense* Belanski, p. 176, Pl. 12, Fig. 1.
1939 *Spinophyllum fasciculare* Soshkina, p. 34, Pl. 7, Figs. 61, 62.
part. 1945 *Disphyllum* (*Synaptophyllum*) cf. *arundinaceum* (Billings); Smith, p. 22 (*non* Billings 1859). (GSC 9330 only, not GSC 9330a = *Thamnophyllum colemanense*).
1945 *Disphyllum* (*Synaptophyllum*) *densum* Smith, p. 22, Pl. 12, Figs. 3a–c.
1951 *Schlüteria fascicularis* (Soshkina); Soshkina, p. 94, Pl. 17, Figs. 1a, b.
1952 *Schlüteria fascicularis* (Soshkina); Soshkina, p. 100, Pl. 40, Fig. 140.
1954 *Schlüteria fascicularis* (Soshkina); Soshkina, p. 44, Pl. 9, Figs. 1–4.
1959 *Phacellophyllum* ? *densum* (Smith); McLaren, p. 30.
? 1970 *Disphyllum fasciculare* (Soshkina); Tsien, p. 172, Figs. 14a, b.
1980 *Disphyllum fasciculare* (Soshkina); Ivanovskiy and Shurygina, p. 25, Pl. 8, Fig. 3.
?*non* 1980 *Peneckiella fascicularis* (Soshkina); Różkowska, p. 21, Pl. 2, Figs. 3, 4.
1980 *Peneckiella szulczewskii* Różkowska, p. 21, Pl. 2, Figs. 2a–e.
? 1981 *Peneckiella fascicularis* (Soshkina); Tsyganko, p. 78, Pl. 24, Figs. 1, 2.
? 1986 *Peneckiella sachaninensis* Lakhov, p. 116, Pl. 1, Figs. 3, 4.
?*non* 1988 *Peneckiella fascicularis* (Soshkina); Rohart, p. 280, Pl. 35, Figs. 10–13.
non 1994 *Peneckiella fascicularis* (Soshkina); Wang, p. 480, Pl. 1, Figs. 1, 2; Pl. 2, Fig. 2.
?*non* 1994 *Peneckiella szulczewskii* Rozkowska; Coen-Aubert, p. 41, Pl. 3, Figs. 7, 8; Pl. 5, Figs. 9–11; Pl. 6, Figs. 3–7.
1995 *Peneckiella fascicularis* (Soshkina); Coen-Aubert, p. 40, Pl. 1, Fig. 1; Pl. 2, Figs. 1–3.
1996 *Peneckiella fascicularis* (Soshkina); Coen-Aubert, p. 63, Pl. 1, Figs. 14–16.
1998 *Disphyllum floydense* (Belanski); Sorauf, p. 59, Pl. 1, Fig. 3; Pl. 27, Figs. 1–5; Pl. 28, Figs. 1–4.
1998 *Peneckiella densa* subsp. A. McLean and Klapper, Pl. 2, Figs. 2, 4.

Material. GSC 6312 (holotype of *Disphyllum* (*Synaptophyllum*) *densum* Smith, 1945); "coral horizon … rapid, 5½ miles above Alexander (sic) Falls", probably small coral bioherm in upper member, Twin Falls Formation, left bank of Hay River, District of Mackenzie, locality 101 (Loc. 3b of Smith 1945, p. 67) of Locality Register and Fig. 5. GSC 126419 (probable topotype of *D.* (*S.*) *densum*); small coral bioherm in upper member, Twin Falls Formation, left bank of Hay River, District of Mackenzie, locality 101 (Amoco Loc. 11140) of Locality Register and Fig. 5. GSC 126420; isolated outcrop, upper member, Twin Falls Formation, right bank of Hay River, District of Mackenzie, locality 112 (Amoco Loc. 11138) of Locality Register and Fig. 5. GSC 126421, 126422; isolated outcrop, upper member, Twin Falls Formation, right bank of Hay River, District of Mackenzie, locality 113 (Amoco Loc. 11139) of Locality Register and Fig. 5. GSC 126423; isolated outcrop, upper member, Twin Falls Formation, left bank of Hay River, District of Mackenzie, locality 114 (Amoco Loc. 11853) of Locality Register and Fig. 5. GSC 126424, 126425; upper member, Twin Falls Formation, left bank of Hay River, immediately upstream from Grumbler Rapids, District of Mackenzie, locality 46 (Amoco Loc. 11143) of Locality Register and Fig. 5. GSC 9330; "coral zone in lower limestone at Grumbler Rapids" (GSC Loc. 5664, Loc. 5 of Smith 1945, p. 67), location uncertain, but possibly same as Amoco Loc. 11143, above. GSC 126426, 126427; member C, Hay River Formation, 5.1–8.3 m above base, 0.0–3.2 m below top, right bank of Hay River, opposite Enterprise, District of Mackenzie, locality 76 (Amoco Loc. 11133) of Locality Register and Fig. 5. GSC 126428; member C, Hay River Formation, 0.3–3.3 m below top, Twin Falls Creek, District of Mackenzie, locality 111 (Amoco Loc. 11136) of Locality Register and Fig. 5. GSC 126429; Mikkwa Formation, 7.5 m above base of 11 m thick section, left bank of Peace River, adjacent to Vermilion Falls, Alberta, locality 15 (Amoco Loc. 14014) of Locality Register and Fig. 6. GSC 126430; limestone equivalent of upper Peechee Member, Southesk Formation, 0.0–26.5 m below base of Cripple Tongue, ridge on northwest side of Cripple Creek, Alberta, locality 64 (Amoco Loc. 11388) of Locality Register and Fig. 11. GSC 126431; Cripple Tongue, 12.8–14.3 m above base, 1.0–2.4 m below top, ridge section approximately 7 km northwest of junction of Ram River and Hummingbird Creek, Alberta, locality 93 (Amoco Loc. 10466) of Locality Register and Fig. 11. GSC 126432; upper Leduc Formation at 3869.2 m in Gulf Shell Colt well, Lsd 5, Sec. 5, Twp 58, Rge 24 W5M, Alberta (Well loc. 10 of Fig. 15).

Population studied. 54 sectioned coralla.

Range. Mid Frasnian, zones 7–11, *M. cameroni* – *P. irregularis* coral faunas.

Diagnosis. Peneckiella with corallite diameter 5–9 mm and 16–25 major septa, which may reach corallite axis, but are more commonly withdrawn to 0.6–0.9 of corallite radius. Peripheral septal dilation is generally moderate to strong and may vary considerably in the one corallite. Dissepiments in 1–4 rows, globose, and peneckielloid, occasionally horseshoe. Tabulae closely spaced, mainly incomplete, in moderately to strongly arched series.

Description. Corallum is phaceloid to rarely cateniform, with peripheral increase and generally closely spaced corallites. Corallite diameter ranges from 5 to 9 mm, with a mean of 7 mm. Major septa vary in number from 16 to 25, most commonly 19–21, with a mean of 19. In many corallites major septa extend to, or almost to, corallite axis, but may be shorter, as little as 0.6 of corallite radius. Minor septa are mainly confined to dissepimentarium. Peripheral fusiform septal dilation is generally moderate to strong. Where strong, septa may be in partial lateral contact, with only intermittent dissepiment development (e.g., GSC 126422, Pl. 18, Figs. 3, 6). In the tabularium, major septa are generally slender and may show some irregular zigzag carinae. Peripherally, both orders of septa commonly show a serrated, "carinate" appearance in transverse section, probably from lateral projections of rhipidacanthine trabeculae (e.g., GSC 126422, Pl. 18, Fig. 3; GSC 126420, Pl. 19, Fig. 1; GSC 126419, Pl. 19, Fig. 6). Trabeculae are generally coarse, in places clearly rhipidacanthine (e.g., GSC 126423, Pl. 20, Fig. 3; GSC 126420, Pl. 20, Fig. 4), and form partial to full fans. Stereome coating of septa peripherally is common.

Dissepiments are small and in 1–4 rows, most commonly 2–3. Globose and peneckielloid forms dominate, but intermittent horseshoes are quite common also, shown in extreme development by GSC 126432 (Pl. 19, Fig. 7). They usually form arched series; more rarely they may locally form axially sloping series of predominantly globose forms. Except in corallites with relatively weak septal dilation, dissepiments typically are coated with stereome. Tabulae are mainly incomplete and closely spaced, commonly 2–3/mm. They characteristically form arched series, although some corallites with shorter septa (e.g., GSC 126430, Pl. 17, Figs. 2, 5) may in places have flat or weakly concave forms.

Remarks. Cylindrophyllum floydense Belanski, 1928 was revised and illustrated in thin section for the first time by Sorauf (1998, pp. 59–61, Pl. 1, Fig. 3; Pl. 27, Figs. 1–5; Pl. 28, Figs. 1–4) and reassigned to *Disphyllum* de Fromentel, 1861. The type material is from the uppermost beds of the Mason City Member, Shell Rock Formation on Shell Rock River, Iowa, and is mid Frasnian, Montagne Noire conodont zone 6–8 in age (Day 1998, Fig. 3). The material illustrated by Sorauf (1998) shows primarily globose dissepiments, with some rare peneckielloid forms. Prof. Sorauf kindly provided a specimen of this species from the type area, which, upon sectioning, revealed well-developed peneckielloid dissepiments (Pl. 18, Fig. 4). One of the sections illustrated by Sorauf (1998, Pl. 27, Fig. 5) shows what appear to be rhipidacanthine trabeculae and this, with the demonstrated presence of peneckielloid dissepiments, leads the present author to consider "*C*". *floydense* is better accommodated in *Peneckiella*. The specimen of *P. floydensis* from Iowa illustrated herein (GSC 126453, Pl. 18, Figs. 4, 5) is typical of many specimens in the Canadian fauna described above, and many coralla in that fauna have corallites with globose dissepiments comparable to those illustrated by Sorauf (1998) in his Iowa collection. The coarse trabeculae, uneven septal dilation in the one corallite, and "carinate" appearance of this dilation, all well illustrated by Sorauf, are characteristically found in the Canadian material. In the past, the Canadian forms have been referred to *Peneckiella densa* (Smith 1945), or variants of that species (McLean and Klapper 1998), but with Sorauf's (1998) revision of "*C*". *floydense* and the evidence of the additional Iowa specimen illustrated herein, the writer has no hesitation in considering the Iowa and Canadian forms as conspecific.

The holotype of *Spinophyllum fasciculare* Soshkina, 1939, from the Frasnian of the western slopes of the southern Urals, was originally illustrated by line drawing (Soshkina 1939, Pl. 7, Figs. 61, 62) and later photographically (Ivanovskiy and Shurygina 1980, Pl. 8, Figs. 3a, b). It was subsequently placed in the genus *Schlueteria* Wedekind, 1921 (a synonym of *Disphyllum* de Fromentel, 1861) by Soshkina (1951, 1952, 1954) and in *Disphyllum* by Ivanovskiy and Shurygina (1980). Later authors have generally referred the species to *Peneckiella* (e.g., Różkowska 1980; Tsyganko 1981; Rohart 1988; Coen-Aubert 1994, 1995, 1996) and the present writer agrees with that assignment. On the basis of the holotype and other material subsequently referred to the species by Soshkina (1951, Pl. 17, Figs. 1a, b; 1954, Pl. 9, Figs. 1–4), *P. fascicularis* has comparable corallite size, septal number, irregular septal dilation in the one corallite, mix of globose, peneckielloid, and horseshoe dissepiments in 1–3 rows, and tabular structure to *P. floydensis*, and the two species are considered here to be synonymous. It is interesting to note that Soshkina (1951, p. 94) recognized the relationship between "*Disphyllum* (*Synaptophyllum*)" *densum* Smith (= *P. floydensis* as interpreted here) and *P. fascicularis* by placing the two in synonymy.

Belgian material assigned to *P. fascicularis* by Coen-Aubert (1995, Pl. 1, Fig. 1; Pl. 2, Figs. 1–3; 1996, Pl. 1, Figs. 14–16; Phillipeville and Grands Breux formations, mid Frasnian) shows no major differences from *P. floydensis* and is referred here to that species. Forms assigned to *P. fascicularis* by other authors, however, are more difficult to classify. Material from the early Frasnian of the Holy Cross Mountains, Poland, referred to *P. fascicularis* by Różkowska (1980, Pl. 2, Figs. 3, 4) appears to have only a single row of dissepiments, and while belonging to *Peneckiella*, is probably not representative of *P. floydensis*. Tsyganko (1981, Pl. 24, Figs. 1, 2) placed early Frasnian material from Pay-Khoy in *P. fascicularis*, and while possibly belonging to that species, and hence *P. floydensis* as here interpreted, the nature of the dissepimentarium is not clear from Tsyganko's illustrations. The material assigned to *P. fascicularis* by Rohart (1988, Pl. 35, Figs. 10–13) from the early Frasnian Ferques Formation of the Boulonnais area, France, appears to have pronounced development of highly arched, mesa-shaped tabulae, and relatively weak septal dilation. While such characters do occur in *P. floydensis*, they are not common and the French material is probably separable from *P. floydensis*. Coen-Aubert (1994) considered Rohart's form to be synonymous with material from the Boussu-en-Fagne Member, Grands Breux Formation (mid Frasnian) of the Dinant Basin, Belgium, that she assigned to *P. szulczewskii* Różkowska, 1980 (Coen-Aubert 1994, p. 41, Pl. 3, Figs. 7, 8; Pl. 5, Figs. 9–11; Pl. 6, Figs. 3–7). Both Rohart's and Coen-Aubert's specimens are likely to be synonymous, but separable from *P. szulczewskii* by having more arched, mesa-shaped tabulae, and more abundant, small, globose dissepiments at the inner margin of the dissepimentarium. The type material of *P. szulczewskii* itself, from the early Frasnian of the Holy Cross Mountains, Poland, does not appear to show any significant differences from *P. floydensis* (Różkowska 1980, Pl. 2, Figs. 2a–e) and is placed here in synonymy with that species. Specimens from the Heyuanzhai Formation (?early Frasnian) of Yunnan, assigned to *P. fascicularis* by Wang (1994, Pl. 1, Figs. 1, 2; Pl. 2, Fig. 2), have axially sloping rows of globose dissepiments and probably belong to *Disphyllum*.

P. sachaninensis Lakhov, 1986 was described from the early Frasnian (Cherkesova 1989) Zhandr Horizon of Novaya Zemlya. It seems closely similar to *P. floydensis*, perhaps differing only in having less dilated septa and a tendency to have fewer rows of dissepiments (Lakhov 1986, Pl. 1, Figs. 3, 4). Without knowledge of its variability, it is regarded for the present as a possible synonym of *P. floydensis*.

Among other Canadian species of *Peneckiella*, *P. floydensis* shows close similarities to *P. metalinae* Sorauf, 1972. The two species overlap in stratigraphic range in western Canada (see Fig. 16), but rarely occur together. *P. metalinae* may be distinguished by its generally smaller corallites, with fewer septa, and usually only one, or at most two rows of dissepiments.

Peneckiella metalinae Sorauf, 1972
Pl. 20, Figs. 5, 6; Pl. 21, Figs. 1–4; Pl. 22, Figs. 1, 2; Pl. 23, Figs. 1–3; Pl. 24, Figs. 1, 3

part. 1945 *Disphyllum* (*Synaptophyllum*) *stramineum* (Billings); Smith, p. 23, Pl. 13, Figs. 6, 9, 11, 12 only (*non* Billings 1859).
1972 *Peneckiella metalinae* Sorauf, p. 431, Pl. 1, Fig. 7; Pl. 2, Figs. 1, 2.
1986 *Peneckiella guangxiense* Yu and Kuang, p. 150, Text-Figs. 8a–d; Pl. 4, Figs. 1–4.
1989 *Peneckiella* sp. McLean and Sorauf, p. 392, Pl. 2, Figs. 5, 8.
1995 *Peneckiella duponti* Coen-Aubert, p. 42, Pl. 2, Figs. 6, 7; Pl. 3, Figs. 1–8.

Material. GSC 126433, 126434; unnamed limestone unit in Imperial Formation, 0.0–6.0 m above base, unnamed tributary of Carlson Creek, District of Mackenzie, locality 2 (Amoco Loc. 11868) of Locality Register and Fig. 3. GSC 126435; Jean-Marie Member, Redknife Formation, 8.0 m below top, middle gorge, Jean-Marie River, District of Mackenzie, locality 37 (Amoco Loc. 11170) of Locality Register and Fig. 4. GSC 126436; Jean-Marie Member, Redknife Formation, 0.0–8.0 m below top, same location as GSC 126435, above. GSC 9341; same horizon and approximately same location (Loc. 21 of Smith 1945, p. 72) as GSC 126435, above. GSC 9337; Jean-Marie Member, Redknife Formation, Table Rock Rapids, Trout River, District of Mackenzie, locality 38 (Loc. 14 of Smith 1945, p. 71) of Locality Register and Fig. 4. GSC

9338; Jean-Marie Member, Redknife Formation, "half a mile below Table Rock Rapid", Trout River, District of Mackenzie, locality 109 (Loc. 13 of Smith 1945, p. 71) of Locality Register and Fig. 4. GSC 126437; talus from upper Mt. Hawk or lower Simla formations, Childear Mountain, Alberta, locality 84 (Amoco Loc. 15251) of Locality Register and Fig. 8. GSC 126438; Mt. Hawk Formation, 100.0 m below top, Winnifred Pass, Alberta, locality 26 (Amoco Loc. 12541) of Locality Register and Fig. 8. GSC 126439; Mt. Hawk Formation, approximately 5 m above base of resistant carbonates, Crowsnest Pass, Alberta, locality 119 (first uncatalogued locality) of Locality Register and Fig. 14. GSC 126440; Zeta Lake Member, Nisku Formation at 2400.2 m, Amoco et al. Bigoray well, Lsd 1, Sec. 7, Twp 52, Rge 8 W5M, Alberta (Well loc. 14 of Fig. 15).

Population studied. 90 sectioned coralla.

Range. Mid – late Frasnian, zones 11–12, ?*Temnophyllum* sp. A – *H. magna* coral faunas.

Diagnosis. Peneckiella with corallite diameter 4–6 mm and 17–20 major septa. Great variation in septal character shown by major septa ranging in length from 0.3 to 0.9 of corallite radius and peripheral dilation being weak to very strong. Dissepiments generally in single row, and peneckielloid, globose, and horseshoe. Tabulae mainly complete and vary from being weakly to strongly arched, including mesa-shaped forms.

Description. Corallum is generally phaceloid, with closely spaced corallites and peripheral increase. Corallite diameter ranges from 4 to 6 mm, with a mean of 5.1 mm. Major septa vary in number from 17 to 20, with a mean of 18. Variation in length of major septa is pronounced, from 0.3 to 0.9 of corallite radius. Such variation can be seen in the one corallum and even in the one corallite (GSC 126436, Pl. 22, Fig. 1). However, a length 0.5–0.7 of radius is most commonly found. Minor septa barely extend into tabularium. Dilation of both orders of septa is also highly variable, ranging from very slight (e.g., GSC 126436, Pl. 22, Fig. 1) to intense peripheral fusiform dilation (e.g., GSC 126435, Pl. 24, Fig. 1). All variation between these extremes may be present. Within the tabularium, major septa are generally very slender and straight. Well-developed trabecular fans occur in peripheral, dilated portions of septa, and in places can be seen to be clearly rhipidacanthine (e.g., GSC 126440, Pl. 23, Fig. 1).

Dissepiments are generally in a single row, peneckielloid, globose, and horseshoe. Most commonly one style predominates in a corallum, but all three are usually present, often in the one corallite. Of the three, horseshoes are the most rarely developed. Stereome coating is not pronounced, but dissepiments may be obscured by trabeculae in longitudinal sections of specimens with strong septal dilation (e.g., GSC 126435, Pl. 24, Fig. 2). Tabulae are mainly complete, almost flat or weakly arched in shorter septate corallites, becoming strongly arched and, in some cases, mesa-shaped in longer septate forms (e.g., GSC 126440, Pl. 23, Figs. 1, 3). Tabulae are fairly evenly and mainly well spaced, usually 1–2/mm.

Remarks. The type material of *P. metalinae* is from an isolated limestone outcrop at Limestone Hill, north of Metaline Falls, Washington (Sorauf 1972). The section at that locality, originally regarded as entirely Middle Devonian (Sorauf 1972, p. 428), has been shown to also include Frasnian strata (McLean and Pedder 1987, p. 158). *Smithiphyllum occidentale* (Sorauf 1972) from the Limestone Hill outcrop is considered to be mid Frasnian in age, on the basis of associated conodonts and foraminifera (McLean and Pedder 1987), and equivalent in age to specimens of *S. occidentale* from Hay River, District of Mackenzie. *P. metalinae* occurs with *S. occidentale* on Hay River, and it seems likely that the type material of *P. metalinae* is also mid Frasnian in age. In western Canada, *P. metalinae* is a particularly widespread species, being common in mid to late (but not latest) Frasnian strata of both outcrop and subsurface, from southern Alberta to the central Mackenzie River valley of District of Mackenzie (Fig. 16).

A number of previously described species of *Peneckiella* show close similarities to *P. metalinae*. In particular, *P. guangxiensis* Yu and Kuang, 1986, from the late Frasnian Rongxian Formation of Guangxi, and *P. duponti* Coen-Aubert, 1995, from the mid Frasnian Lion Member, Grands Breux Formation of the Dinant Basin, Belgium, show no significant differences from *P. metalinae* and are placed here in synonymy with that species. The type species, *P. minor* (Roemer 1855) from the mid Frasnian Ibergerkalk of the Harz Mountains, Germany, is comparable to *P. metalinae* in corallite size and septal number. Based on the author's topotypic material, illustrated in part herein (Pl. 15, Fig. 7; Pl. 16, Figs. 1–6; Pl. 17, Fig. 1), it may be separated by having a greater tendency to horseshoe development and generally longer major septa. However, the two species are closely similar. *P. salternensis* Scrutton, 1968 from the Frasnian of Saltern Cove, Devon, is unfortunately poorly preserved and the dissepiment structure, in particular, is difficult to determine from the illustrations (Scrutton 1968, Pl. 18, Figs. 1–4). It appears to differ from *P. metalinae* by having greater horseshoe development (although these are not clear in the illustrations) and having very closely spaced, regularly arched tabulae.

Considering the most similar Canadian representatives of *Peneckiella*, *P. metalinae* may be distinguished from

P. floydensis (Belanski, 1928) by having smaller corallites, fewer septa, and usually only a single row of dissepiments. *P. metalinae* is separated from *P. gracilis* n. sp. by having larger corallites, more septa, generally shorter major septa, and usually less arched tabulae.

Peneckiella gracilis n. sp.
Pl. 22, Fig. 3; Pl. 23, Figs. 4, 5; Pl. 24, Figs. 2, 4

Derivation of name. Latin, gracilis = slender, a reference to the small corallites of this species.

Type material. GSC 126441 (holotype), 126442, 126443 (paratypes); Cripple Tongue, 24.4–33.5 m above base, 32.3–41.4 m below top, south side of north branch of Hummingbird Creek, Ram Range, Alberta, locality 118 (Amoco Loc. 11495) of Locality Register and Fig. 11. GSC 15481, 15482 (paratypes); probably same horizon and location as Amoco Loc. 11495, above, locality 118 (uncatalogued) of Locality Register and Fig. 11. GSC 126444, 126445 (paratypes), uppermost Peechee Member, Southesk Formation, 123.3–130.0 m above base, 0.0–6.7 m below top, ridge on north side of main branch of Hummingbird Creek, Ram Range, Alberta, locality 79 (Amoco Loc. 11402) of Locality Register and Fig. 11.

Population studied. 7 sectioned coralla.

Range. Mid Frasnian, zone 11, *P. irregularis* coral fauna.

Diagnosis. Peneckiella with corallite diameter 3–4 mm and 14–16 major septa, which are long, extending to, or slightly withdrawn from corallite axis. Dissepiments usually in single row, peneckielloid, and globose, with horseshoes very rare. Tabulae complete and incomplete, closely spaced, most commonly arched, and in some cases mesa-shaped.

Description. Corallum is phaceloid to dendroid, with corallites mainly closely spaced and having peripheral increase. Corallite diameter varies from 3 to 4 mm, with a mean of 3.4 mm. Major septa are 14–16 in number in mature corallites (mean 15), typically reaching corallite axis or slightly withdrawn. Minor septa are generally confined to the dissepimentarium. Both orders of septa show moderate to strong peripheral fusiform dilation, with major septa being very slender in the tabularium. Trabecular fans are present in the peripheral, dilated parts of the septa, and in some cases rhipidacanthine trabeculae can be discerned (e.g., GSC 126448, Pl. 26, Fig. 5).

Dissepiments generally form a single row, are small and primarily peneckielloid and globose. Very small, isolated horseshoe dissepiments occur rarely. Stereome coating of dissepiments is minor. Tabulae are closely spaced (generally 2–3/mm), complete and incomplete, and most commonly form arched series. In some cases, mesa-shaped tabulae are present (e.g., holotype, GSC 126441, Pl. 24, Fig. 4).

Remarks. This species is morphologically very similar to *Peneckiella metalinae* Sorauf, 1972, common in the western Canadian mid to late Frasnian (see above). It may be distinguished from the latter by having consistently smaller corallites, fewer septa, and usually more arched tabulae. *P. haultainensis* n. sp., from younger Frasnian strata in western Canada (see below), is comparable to *P. gracilis* in corallite size and septal number, but can be separated by having shorter major septa and more flattened tabulae.

The only previously described species with close similarities to *P. gracilis* appears to be *P. liujingensis* Yu and Kuang, 1986, from the late Frasnian Rongxian Formation of Guangxi. This species has comparable corallite dimensions and septal number to *P. gracilis*, but seems to have shorter major septa and considerably less arched tabulae than are typical for that species (Yu and Kuang 1986, Fig. 9, Pl. 4, Figs. 5, 6).

Peneckiella haultainensis n. sp.
Pl. 25, Figs. 1–8; Pl. 26, Figs. 1–6

Derivation of name. After the type locality, Mt. Haultain, Alberta.

Type material. GSC 126446 (holotype), 126447 (paratype); Simla Formation, 9.2 m below top, southeast face of Mt. Haultain, Alberta, locality 55 (R. K. Jull Loc. L-272) of Locality Register and Fig. 9. GSC 126448, 126449 (paratypes); Simla Formation, 80.5–81.0 m below top, south side of Surprise Pass, British Columbia, locality 25 (Amoco Loc. 11513) of Locality Register and Fig. 8. GSC 126450 (paratype); upper Mt. Hawk Formation, 40.0–45.0 m below base of Palliser Formation, Whitehorse Creek tributary, Miette Range, Alberta, locality 78 (Amoco Loc. 10126) of Locality Register and Fig. 10. GSC 126451 (paratype); upper Mt. Hawk Formation, 99.75–101.75 m below base of Palliser Formation, same location as GSC 126450, above. GSC 126452 (paratype); Simla Formation, 21.5 m above base, 57.0 m below top, northwest flank of Mt. Mackenzie, Alberta, locality 28 (Amoco Loc. 11411) of Locality Register and Fig. 10.

Population studied. 9 sectioned coralla.

Range. Latest Frasnian, zone 13, *P. variabilis – S. cinctus* coral faunas.

Diagnosis. Peneckiella with corallite diameter 3.0–3.8 mm and 14–17 major septa, which are short, generally extending only slightly into tabularium. Dissepiments are typically in single row, small, mainly peneckielloid and globose, with occasional horseshoes. Tabulae are mainly complete, closely spaced, weakly arched, flat, or slightly concave.

Description. Corallum is dendroid to phaceloid, with generally well-spaced corallites and peripheral increase. Corallite diameter ranges from 3.0 to 3.8 mm, with a mean of 3.3 mm. Major septa are 14–17 in number (mean 15), mainly short, barely extending into tabularium. In rare cases, major septa may extend up to about 0.7 of corallite radius, but their length in such specimens varies greatly within the one corallum (e.g., GSC 126446, holotype, Pl. 25, Fig. 2) and is still predominantly short. Minor septa are confined to the dissepimentarium. Both orders of septa exhibit peripheral fusiform dilation, generally weak, but moderate in some corallites. Well-developed trabecular fans are present in dilated portions of septa.

Dissepiments are mainly in a single row, rarely in two. They are small, peneckielloid, and globose, with isolated horseshoe dissepiments occasionally developed. Stereome coating is generally weak or absent, although dissepiments may be obscured by trabeculae in longitudinal sections. Tabulae are mainly complete and closely spaced, generally 2–4/mm; they may be flat, weakly arched, or slightly concave.

Remarks. This is the youngest species of *Peneckiella* in western Canada, occurring in beds of late Frasnian age (equivalent to conodont zone 13). Those beds also commonly contain *Thamnophyllum tructense* (McLaren 1959), described above, and that species is generally indistinguishable from *P. haultainensis* in transverse section. Differences lie, however, in the character of the dissepimentarium. In *T. tructense*, there is typically a continuous pipe of horseshoe dissepiments, although very rarely some of these may "unroll" to give a peneckielloid appearance. *P. haultainensis* has only rare, isolated horseshoes, with globose and peneckielloid dissepiments dominating.

P. haultainensis is comparable to *P. gracilis* n. sp. in corallite size and septal number. The latter may be distinguished, however, by having longer major septa and a generally more complex tabularium, with arched and often incomplete tabulae. *P. lateseptata* (Różkowska 1960) from the late Frasnian of Mokrzeszów, Sudeten Mountains, Poland, is also comparable to *P. haultainensis* in corallite size and septal number, and additionally has very short major septa as in the Canadian form. It may be separated from *P. haultainensis* by having more complete and irregularly spaced tabulae, together with weaker peneckielloid dissepiment development (Różkowska 1960, Figs. 31–33). *P. liujingensis* Yu and Kuang, 1986, from the late Frasnian Rongxian Formation, Guangxi, has small corallites as in *P. haultainensis*, with similar development of dissepiments and tabulae. However, *P. liujingensis* appears to have generally longer major septa than *P. haultainensis* (Yu and Kuang 1986, Fig. 9, Pl. 4, Figs. 5, 6).

Locality Register

Localities are listed in order of numbered locations on locality maps, Figs. 3–14. Localities listed below that were also cited by McLean and Pedder (1984, 1987) and McLean (1994*a*, *b*) have the same numbers in each publication.

2. Amoco Loc. 11868, unnamed limestone unit in Imperial Formation, 0.0–6.0 m above base of limestone unit. Right side of unnamed tributary of Carlson Creek, District of Mackenzie. 62°38′06″N, 123°47′58″W. Collected by R.A. McLean, 1981.

3. Amoco Loc. 11163, Kakisa Formation, 44.8 m below top, near lowest falls in Trout River gorge; 24.6–25.9 m below top, immediately below Whittaker Falls, Trout River, District of Mackenzie. 61°08′23″N, 119°50′17″W. Collected by R.A. McLean, 1980.

   Loc. 15 of Smith (1945, p. 71), "conspicuous coral horizon with marcasite nodules near top of cliff" Trout River. Equivalent to beds 24.6–25.9 m below top of Kakisa Formation, above. Collected by E.J. Whittaker, 1921.

13. Amoco Loc. 11210, member C, Hay River Formation, 7.0–12.0 m above base, 0.0–5.0 m below top of member. Left bank of Hay River, immediarely east of village of Enterprise, District of Mackenzie. 60°33′15″N, 116°08′ 20″W. Collected by R.A. McLean, 1979, 1980.

14. Loc. 1a of Smith (1945, p. 67), "lowest beds in cliff…5 miles below foot of portage around Alexander (sic) Falls", ? member B or C, Hay River Formation. Hay River, District of Mackenzie (precise location uncertain). Collected by E.J. Whittaker, 1917.

    Loc. 1b of Smith (1945, p. 67), "beds between 120 and 140 feet elevation". Other location and collection data as for Loc. 1a, above.

    Loc. 1c of Smith (1945, p. 67), "highest beds in cliff", ? member C, Hay River Formation. Other location and collection data as for Loc. 1a, above.

15. Amoco Loc. 14014, Mikkwa Formation, 7.5 m and 7.5–9.75 m above base of 11 m thick exposed section. Left bank of Peace River, adjacent to Vermilion Falls, Alberta. 58°22′26″N, 114°52′03″W. Collected by R.A. McLean, 1985.

25. Amoco Loc. 11513, Simla Formation, 80.5–81.0 m below top. South side of Surprise Pass, between Wallbridge and Bastille Mountains, British Columbia. 53°52′22″N, 120°02′15″W. Collected by R.A. McLean, 1981.

26. Amoco Loc. 12541, Mt. Hawk Formation, 100.0 m, 42.4–45.5 m and 35.4 m below top. Winnifred Pass, Alberta. 53°38′40″N, 119°11′30″W. Collected by R.A. McLean, 1979.

27. R.K. Jull Loc. L-199, upper Southesk Formation, 2.4 m thick unit of thickly bedded limestone, approximately 12.2 m below base of Simla Formation. West side of Mt. Haultain cirque, Alberta. 53°12′N, 118°18′W. Collected by R.K. Jull, 1972.

28. Amoco Loc. 11411, Simla Formation, 21.5 m above base, 57.0 m below top. Northwest flank of Mt. Mackenzie, Alberta. 52°51′N, 117°16′W. Collected by A.S. Hedinger, 1980.

Amoco Loc. 11496, Simla Formation, 69.8–76.5 m above base, 0.0–6.7 m below top. Approximately same location as Amoco Loc. 11411, above. Collected by R.H. Workum, 1979.

37. Amoco Loc. 11170, Jean-Marie Member, Redknife Formation, 8.0 m below top and 0.0–8.0 m below top. Middle gorge, Jean-Marie River, District of Mackenzie. 61°20′06″N, 121°05′53″W. Collected by R.A. McLean, 1980.

Loc. 21 of Smith (1945, p. 72), same horizon and approximately same location as Amoco Loc. 11170 above. Collected by E.J. Whittaker, 1922.

38. Amoco Loc. 11161, Jean-Marie Member, Redknife Formation, 0.0–2.5 m below top. Table Rock Rapids, Trout River, District of Mackenzie. 61°13′27″N, 119°54′48″W. Collected by R.A. McLean, 1980.

Amoco Loc. 15683, same horizon and location as Amoco Loc. 11161, above. Collected by R.A. McLean, 1990.

Loc. 14 of Smith (1945, p. 71), same horizon and location as Amoco Loc. 11161, above. Collected by E.J. Whittaker, 1921.

39. Amoco Loc. 11155, upper member, Redknife Formation, isolated outcrop of 5 m thickness. Bouvier River, District of Mackenzie. 61°08′55″N, 119°01′19″W. Collected by R.A. McLean, 1980.

45. GSC Loc. 30444, upper member, Twin Falls Formation. Right bank of Hay River, approximately 1.6 km below Grumbler Rapids, Hay River, District of Mackenzie. Approximately 60°14′28″N, 116°35′17″W. Collected by D.J. McLaren, 1957.

46. Amoco Loc. 11143, upper member, Twin Falls Formation. Left bank of Hay River, immediately upstream from Grumbler Rapids, District of Mackenzie. 60°13′23″N, 116°36′23″W. Collected by R.A. McLean, 1979.

GSC Loc. 5664, "coral zone in lower limestone at Grumbler Rapids" (Loc. 5 of Smith 1945, p. 67). Exact location unknown, but possibly same as Amoco Loc. 11143 above. Collected by E.J. Whittaker, 1917.

47. GSC Loc. 30442, upper member, Twin Falls Formation. Right bank of Hay River at Grumbler Rapids, District of Mackenzie. 60°13′27″N, 116°35′17″W. Collected by D.J. McLaren, 1957.

55. R.K. Jull Loc. L-272, Simla Formation, 9.2 m below top. Southeast face of Mt. Haultain, Alberta. 53°11′05″N, 118°16′07″W. Collected by R.K. Jull, 1972.

R.K. Jull Loc. L-218, Mt. Hawk Formation, immediately overlying Southesk Formation. Same location and collection data as L-272, above.

R.K. Jull Loc. L-223, Southesk Formation. Same location and collection data as L-272, above.

59. Amoco Loc. 11423, undifferentiated Southesk Formation, upper Mt. Hawk Formation equivalent, 96.9–106.1 m below base of Palliser Formation. Ridge northwest of Mt. Gregg, Alberta. 53°02′26″N, 117°29′08″W. Collected by R.H. Workum, 1980.

63. Amoco Loc. 11416, Cripple Tongue, 33.8 m above base, 52.1 m below top. Ridge northeast of small lake near headwaters of Allstones Creek, Alberta. 52°16′05″N, 116°28′29″W. Collected by A.S. Hedinger, 1980.

64. Amoco Loc. 11388, limestone equivalent of upper Peechee Member, Southesk Formation, 0.0–26.5 m below base of Cripple Tongue. Ridge on northwest side of Cripple Creek, Alberta. 52°10′N, 116°05′52″W. Collected by A.S. Hedinger, 1980.

66. Amoco Loc. 11522, upper member, Cairn Formation, talus from basal 32.3 m. Ridge section, 4 km west of North Burnt Timber Creek, Alberta. 51°29′30″N, 115°28′W. Collected by A.S. Hedinger, 1980.

67. Amoco Loc. 11520, Cairn Formation, 60.7 m below top. Eastern flank of Mt. McDougall, Alberta. 50°53′39″N, 115°02′37″W. Collected by A.S. Hedinger, 1980.

71. Amoco Loc. 15261 (= Amoco Loc. 15447 of McLean 1994*b*, p. 89), Simla Formation, 0.0–5.0 m above base. Eastern flank of Bastille Moun-

tain, British Columbia. 53°52′22″N, 120°00′56″W. Collected by R.A. McLean, 1988.

74. Loc. 20 of Smith (1945, p. 71), upper Kakisa Formation, "bed b, a dolomitic limestone with a few corals" (= bed m of McLaren 1959, p. 30). Coral Falls, Trout River, District of Mackenzie. 61°08′06″N, 119°49′43″W. Collected by E.J. Whittaker, 1921.

75. Uncatalogued locality, ? member B or C, Hay River Formation. Hay River, "forty miles above its mouth" (Whiteaves 1891), precise location unknown, but probably between localities 13 and 75, Fig. 5. Collected by R.G. McConnell, 1887.

76. Amoco Loc. 11133, member C, Hay River Formation, 5.1–8.3 m above base, 0.0–3.2 m below top. Right bank of Hay River, opposite village of Enterprise, District of Mackenzie. 60°33′06″N, 116°07′47″W. Collected by R.A. McLean, 1980.

78. Amoco Loc. 10126, upper Mt. Hawk Formation, equivalent to Simla Formation, 40.0–45.0 m below base of Palliser Formation. Tributary of Whitehorse Creek, Miette Range, Alberta. 52°59′07″N, 117°32′20″W. Collected by R.A. McLean, 1979.

Amoco Loc. 15269, same unit as above, 99.75–101.75 m below base of Palliser Formation. Same location as above. Collected by R.A. McLean, 1983.

79. Amoco Loc. 11402, Peechee Member, Southesk Formation, 0.0–6.7 m below top, 123.3–130.0 m above base. Ridge on north side of main branch of Hummingbird Creek, Ram Range, Alberta. 52°02′30″N, 116°11′W. Collected by A.S. Hedinger, 1980.

84. Amoco Loc. 15251, talus from upper Mt. Hawk or lower Simla formations. Eastern flank of Childear Mountain, Alberta. 53°39′53″N, 119°17′03″W. Collected by R.A. McLean, 1988.

85. Amoco Loc. 15253, upper Southesk Formation, 25.0–36.0 m below base of Simla Formation. Eastern side of valley cut by South Berland River, Persimmon Range, Alberta. 53°32′38″N, 118°40′39″W. Collected by R.A. McLean, 1988.

93. Amoco Loc. 10466, Cripple Tongue, 1.0–2.4 m below top, 12.8–14.3 m above base. Ridge section, approximately 7 km northwest of junction of Ram River and Hummingbird Creek, Alberta. 52°07′20″N, 116°00′15″W. Collected by A.S. Hedinger, 1979.

95. Amoco Loc. 11175, Jean-Marie Member, Redknife Formation, 9.3–12.0 m below top, 0.0–2.7 m above base. Birch River, District of Mackenzie. 61°16′N, 122°01′30″W. Collected by R.A. McLean, 1980.

99. Amoco Loc. 11146, Alexandra Member, Twin Falls Formation, 0.0–4.3 m above base. North side of McNallie Creek, immediately east of Mackenzie Highway crossing, below falls, District of Mackenzie. 60°46′53″N, 116°34′31″W. Collected by R.A. McLean, 1979.

100. Amoco Loc. 11131, member B, Hay River Formation, 1.95–22.65 m above base. Left bank of Hay River, District of Mackenzie. 60°33′31″N, 116°06′23″W. Collected by R.A. McLean, 1980.

101. Amoco Loc. 11140, bioherm in upper member, Twin Falls Formation. Left bank of Hay River, District of Mackenzie. 60°27′16″N, 116°19′43″W. Collected by R.A. McLean, 1979.

Loc. 3b of Smith (1945, p. 67), "coral horizon .... rapid, 5½ miles above Alexander (sic) Falls", Hay River, District of Mackenzie. Probably same horizon and location as Amoco Loc. 11140, above. Collected by E.J. Whittaker, 1917.

104. Amoco Loc. 15254, upper Southesk Formation, 5.0–6.0 m below contact with upper Mt. Hawk Formation. Ridge approximately 5 km west of junction of Wildhay River and Eagle's Nest Creek, Persimmon Range, Alberta. 53°30′26″N, 118°35′28″W. Collected by R.A. McLean, 1988.

109. Amoco Loc. 15670, Jean-Marie Member, Redknife Formation, 0.0–3.75 m below top and above base. Left bank of Trout River, downstream from Table Rock Rapids, District of Mackenzie. 61°14′39″N, 119°54′43″W. Collected by R.A. McLean, 1990.

Loc. 13 of Smith (1945, p. 71), "half a mile below Table Rock Rapid", Trout River. Same horizon and probably same location as Amoco Loc. 15670, above. Collected by E.J. Whittaker, 1921.

110. Amoco Loc. 14302, member B, Hay River Formation. Isolated outcrop in gravel pit beside Hay River Highway, District of Mackenzie. 60°37′18″N, 116°04′11″W. Collected by R.A. McLean, 1986.

111. Amoco Loc. 11136, member C, Hay River Formation, 0.0–3.3 m below top. Twin Falls (Escarpment) Creek, District of Mackenzie. 60°31′29″N, 116°12′30″W. Collected by R.A. McLean, 1979.

112. Amoco Loc. 11138, upper member, Twin Falls Formation, isolated 8 m thick outcrop, 4.3 m above base. Right bank of Hay River, District of Mackenzie. 60°28′39″N, 116°18′37″W. Collected by R.A. McLean, 1980.

113. Amoco Loc. 11139, upper member, Twin Falls Formation, isolated 3.2 m thick outcrop, 2.5 m above base. Right bank of Hay River, District of Mackenzie. 60°27′50″N, 116°19′43″W. Collected by R.A. McLean, 1980.

114. Amoco Loc. 11853, upper member, Twin Falls Formation, isolated outcrop. Left bank of Hay River, District of Mackenzie. 60°25′28″N, 116° 20′W. Collected by R.A. McLean, 1981.
115. Amoco Loc. 14055, Kakisa Formation, 13.0–18.0 m above base, 20.0–25.0 m below top. "Ottertail Ridge", approximately 19 km west of junction of Nabesche River and Emerslund Creek, British Columbia. 56°15′49″N, 123°26′13″W. Collected by R.A. McLean, 1985.
116. Amoco Loc. 15252, Simla Formation, 40.4–65.4 m above base, 0.0–25.0 m below top. Persimmon Creek, Persimmon Range, Alberta. 53°33′34″N, 118°43′26″W. Collected by R.A. McLean, 1988.
117. GSC Loc. 36863, Mt. Hawk Formation, 113.4–115.5 m above base, 8.8–11.0 m below top (unit 26 of Mountjoy 1965, p. 53). Roche Miette, Alberta. 53°10′15″N, 117°55′W. Collected by E.W. Mountjoy, 1958.
118. Amoco Loc. 11495, Cripple Tongue, 24.4–33.5 m above base, 32.3–41.4 m below top. South side of north branch of Hummingbird Creek, Ram Range, Alberta. 52°03′30″N, 116°14′W. Collected by R.A. McLean, 1994.

    Uncatalogued locality, Cripple Tongue. Probably same horizon and location as Amoco Loc. 11495, above. Collected by C.W. Stearn, 1954.
119. Uncatalogued locality, Mt. Hawk Formation, approximately 5 m above base of lowest resistant, argillaceous limestones. Crowsnest Pass, railway cut adjacent to Crowsnest Lake, Alberta. 49°38′04″N, 114°38′24″W. Collected by R.A. McLean, 1993.

    Uncatalogued locality, Mt. Hawk Formation, 173.7–176.8 m below base of Palliser Formation, north side of Crowsnest Pass. Probably the same horizon and location as the uncatalogued locality, above. Collected by A.E.H. Pedder, 1959.

    Uncatalogued locality, "basal beds of the Crowsnest section" (Warren 1928, p. 112). Precise horizon and location unknown, but possibly the same as the uncatalogued locality, above. Collected by P.S. Warren, 1927.
120. Uncatalogued locality, Alexandra Member, Twin Falls Formation, fore-reef debris facies. Isolated roadcut exposing 3 m thick section, west side of Mackenzie Highway near Heart Lake, District of Mackenzie. 60°50′16″N, 116°36′54″W. Collected by R.A. McLean, 2001.
121. GSC Loc. 25176, Cripple Tongue ("grey mudstone and limestone member, Southesk Formation" of McLaren 1956, pp. 52–53), 8.0–30.5 m above base, 7.9–20.4 m below top (unit 22 of McLaren 1956, p. 53). Job Creek headwaters, Whiterabbit Range, Alberta (Section 23 of McLaren 1956). Approximately 52°20′N, 116°49′W. Collected by D.J. McLaren, 1953.

References

Assereto, R. 1962. Coelenterati devonici del Monte Osternig (Alpi Carniche). Rivista Italiana di Paleontologia, **68**: 3–39.

Aung, A.K. 1995. New Middle Devonian (Eifelian) rugose corals from Myanmar. Journal of Southeast Asian Earth Sciences, **11**: 23–32.

Belanski, C.H. 1928. Descriptions of some typical fossils of the Shellrock Stage. American Midland Naturalist, **11**: 171–212.

Bertrand, M., Coen-Aubert, M., Dumoulin, V., Préat, A., and Tourneur, F. 1993. Sédimentologie et paléoécologie de l'Emsien supérieur et de l'Eifelien inférieur des regions de Couvin et de Villers-la-Tour (bord sud du Synclinorium de Dinant, Belgique). Neues Jahrbuch für Geologie und Paläontologie Abhandlungen, **188**: 177–211.

Besprozvannykh, N.I., Dubatolov, V.N., Kravtsov, A.G., Latypov, Yu. Ya., and Spasskiy, N. Ya. 1975. Devonskie rugozy Taymyro-Kolymskoy provintsii. Institut Geologii i Geofiziki Sibirskoe Otdelenie Akadamiya Nauk SSSR, Trudy 228.

Billings, E. 1859. On the fossil corals of the Devonian rocks of Canada West. Canadian Journal of Industry, Science and Art, **4**: 97–140.

Birenheide, R. 1969. Typen mittel- und oberdevonischer Rugosa aus der Sammlung Goldfuss. Senckenbergiana Lethaea, **50**: 37–55.

Birenheide, R. 1978. Leitfossilien no. 2. Rugose Korallen des Devon. Gebrüder Borntraeger, Berlin, Stuttgart.

Birenheide, R., and Liao, Wei-hua. 1985. Rugose Korallen aus dem Givetium von Dushan, Provinz Guizhou, S-China. 3: Einzelkorallen und einige Koloniebildner. Senckenbergiana Lethaea, **66**: 217–267.

Birenheide, R., and Soto, F.M. 1992. Rugose Einzel- und Phaceloid-Korallen aus dem Ober-Givetium (Mittel-Devon) des Kantabrischen Gebirges, NW-Spanien. Palaeontographica, Abteilung A, **221**: 95–123.

Brice, D. 1970. Étude paléontologique et stratigraphique du Dévonien de l'Afghanistan. Contribution a la connaissance des brachiopods et des polypiers rugeux. Notes et Mémoires sur le Moyen-Orient, **11**: 1–364.

Brice, D., and Rohart, J.-C. 1974. Les Phillipsastraeidae (Rugosa) du Dévonien de Ferques (Boulonnais, France). Première note. Le genre *Macgeea* Webster, 1889. Nouvelles observations. Annales de la Société Géologique du Nord, **94**: 47–62, Pls. 7–9.

Bulvanker, E.Z. 1958. Devonskie chetyrekhluchevye korally okrani Kuznetskogo Basseyna. Vsesoyuzniy Nauchno-Issledovatelskiy Geologicheskiy Institut (VSEGEI), Leningrad. Text and Atlas.

Cai, Tu-ci, and Zeng, Cai-lin. 1983. Coelenterata. *In* Atlas of paleontology of N.W. China. Xinjiang (Sinkiang) volume, Part II. Geological Publishing House, Beijing, pp. 113–216, 684–704, Pls. 31–65. (In Chinese).

Cao, Xuan-duo, and Ouyang, Xuan. 1987. Late Silurian and Devonian corals from Luqu and Tewo regions, west Qinling Mountains. *In* Late Silurian-Devonian strata and fossils from Luqu-Tewo area of west Qinling Mountains, China, Volume 1. Nanjing University Press, Nanjing, pp. 139–202, Pls. 14–37. (In Chinese with English abstract).

Cao, Xuan-duo, Ouyang, Xuan, and Jin, Tong-an. 1982. Rugosa. *In* Paleontological atlas of northwest China. Shaanxi, Gansu and Ningxia volume. Part II, Upper Paleozoic. *Edited by* Xian Institute of Geology and Mineral Resources. Geological Publishing House, Beijing, pp. 46–179, 565–593, Pls. 15–60. (In Chinese).

Cherkesova, S.V. 1989. Lower and Middle Devonian marine deposits of the Soviet Arctic and the correlation with Arctic Canada. *In* Devonian of the world. *Edited by* N.J. McMillan, A.F. Embry, and D.J. Glass. Canadian Society of Petroleum Geologists, Memoir 14, Volume 3, pp. 669–679. (Imprint 1988).

Coen-Aubert, M. 1982. Rugueux solitaires du Frasnien de la Belgique. Bulletin de l'Institut Royal des Sciences Naturelles de Belgique. Sciences de la Terre, **54**: 1–65.

Coen-Aubert, M. 1994. Stratigraphie et systématique des Rugueux de la partie moyenne du Frasnien de Frasnes-lez-Couvin (Belgique). Bulletin de l'Institut Royal des Sciences Naturelles de Belgique. Sciences de la Terre, **64**: 21–56.

Coen-Aubert, M. 1995. Espèces du genre *Peneckiella* Soshkina, 1939 dans le Frasnien de la Belgique. Bulletin de l'Institut Royal des Sciences Naturelles de Belgique. Sciences de la Terre, **65**: 35–49.

Coen-Aubert, M. 1996. Rugueux Frasnien du sondage de Focant. Annales de la Société Géologique de Belgique, **117**: 57–67.

Coen-Aubert, M. 1998. Thamnophyllides et Acanthophyllides près de la limite Eifelien-Givetien à Wellin et Pondrôme (Belgique). Bulletin de l'Institut Royal des Sciences Naturelles de Belgique. Sciences de la Terre, **68**: 5–24.

Coen-Aubert, M., and Wrzołek, T. 1991. Redescription of the rugose coral *Macgeea* (*Rozkowskaella*) *sandaliformis* (Różkowska 1980) from the Upper Frasnian of the Holy Cross Mountains (Poland). Bulletin de l'Institut Royal des Sciences Naturelles de Belgique. Sciences de la Terre, **61**: 5–19.

Cotton, G. 1973. The rugose coral genera. Elsevier, Amsterdam.

Crickmay, C.H. 1962. New Devonian fossils from western Canada. Published by the author, Calgary.

Dames, W. 1869. Herr Dames an Herrn Kunth. Zeitschrift der Deutschen geologischen Gesellschaft, **21**: 699–700.

Day, J. 1998. Distribution of latest Givetian-Frasnian Atrypida (Brachiopoda) in central and western North America. Acta Palaeontologica Polonica, **43**: 205–240.

de Blainville, H.M.D. 1830. Zoophytes. *In* Dictionnaire des Sciences Naturelles, 60. *Edited by* G.F. Cuvier. F.G. Levrault, Paris. (Not seen).

de Wit, R., and McLaren, D.J. 1950. Devonian sections in the Rocky Mountains between Crowsnest Pass and Jasper, Alberta. Geological Survey of Canada, Paper 50–23.

Douglas, R.J.W., and Norris, D.K. 1976. Geology, Camsell Bend, District of Mackenzie. Geological Survey of Canada, Map 1375A, scale 1 : 250 000.

Dubatolov, V.N., and Spasskiy, N. Ya. 1971. Devonskie korally Dzhungaro-Balkhashskoy provintsii. Institut Geologii i Geofiziki Sibirskoe Otdelenie Akadamiya Nauk SSSR, Trudy 74.

Dybowski, W. 1873–1874. Monographie der Zoantharia sclerodermata rugosa aus der Silurformation Estlands, Nord Livlands und der Insel Gotland. Archiv für die Naturkunde Liv-, Ehst- und Kurlands, **5**: 257–414, Pls. 1, 2 (1873), 415–532, Pls. 3–5 (1874).

Ehrenberg, C.G. 1834. Beiträge zur physiologischen Kenntniss der Corallenthiere im allgemeinen, und besonders des rothen Meeres, nebst einem Versuche zur physiologischen Systematik derselben. Abhandlungen der Königlichen Akademie der Wissenschaften zu Berlin (1832), pp. 225–380.

Ermakova, K.A. 1957. Novye vidy devonskikh korallov rugoza iz tsentralnykh oblastey Russkoy Platformy. Vsesoyuznyy Nauchno-Issledovatelskiy Geologo-Razvedochnyy Neftyanoy Institut, Trudy 8, pp. 160–191.

Errenst, C. 1993. Koloniebildende Phillipsastreidae und Hexagonariinae aus dem Givetium des Messinghäuser Sattels und vom Südrand des Briloner Massenkalkes (nordöstliches Sauerland). Geologie und Paläontologie in Westfalen, **26**: 7–45.

Etheridge, R., Jr. 1890. On the occurrence of the genus *Tryplasma* Lonsdale (*Pholidophyllum* Lindström) and another coral apparently referrable to *Diphyphyllum* Lonsdale in the Upper Silurian and Devonian rocks, respectively, of N.S. Wales. Records of the Geological Survey of New South Wales, **2**: 15–20, Pl. 1.

Etheridge, R., Jr. 1899. Descriptions of new or little-known Victorian Palaeozoic and Mesozoic fossils, no. 1. Geological Survey of Victoria, Progress Report 11, pp. 30–36, Pls. A, B.

Exploration Staff, Chevron Standard Limited 1979. The geology, geophysics and significance of the Nisku reef discoveries, West Pembina area, Alberta, Canada. Bulletin of Canadian Petroleum Geology, **27**: 326–359.

Fedorowski, J. 1968. *Pterorrhiza densa* n. sp. (Tetracoralla) from the Lower Givetian of Skaly (Holy Cross Mountains). Rocznik Polskiego Towarzystwa Geologicznego, **38**: 225–228, Pl. 21.

Fedorowski, J. 2003. Podgromady: Rugosa Milne Edwards et Haime, 1850, Dividocorallia Fedorowski, 1991. *In* Budowa geologiczna Polski. Tom III. Atlas skamieniałosci przewodnich i charakterystycznych, część 1b, Dewon. *Edited by* L. Malinowska. Państwowy Instytut Geologiczny, Warszawa, pp. 49–124, 797–808, Pls. 14–66.

Flügel, H.W. 1956. Kritische Bemerkung zum Genus *Peneckiella* Soshkina. Neues Jahrbuch für Geologie und Paläontologie Monatshefte 1956: pp. 355–365.

Flügel, H.W., and Hubmann, B. 1994. Catalogus Fossilium Austriae. Heft IVc/1a. Anthozoa palaeozoica: Rugosa. Österreichischen Akadamie der Wissenschaften, Wien.

Frech, F. 1885. Die Korallenfauna des Oberdevons in Deutschland. Zeitschrift der Deutschen geologischen Gesellschaft, **37**: 21–130, Pls. 1–11.

Frech, F. 1886. Die Cyathophylliden und Zaphrentiden des deutschen Mitteldevon. Paläontologische Abhandlungen, **3**: 115–234, Pls. 13–20.

Frech, F. 1900. Zur Kenntnis der mittleren Paläozoicum in Hocharmenien und Persien. Paläontologisches Institut der Universität Wien, Mittheilungen 12, pp. 183–208, Pls. 15–17.

Galle, A. 1993. Middle Devonian Rugosa from Hornú Benešov (Moravia, Czech Republic). Journal of the Czech Geological Society, **38**: 59–70.

Gischler, E., Weller, H., and Weyer, D. 1991. Devonian reefs of the Harz Mountains, Germany. Sixth International Symposium on Fossil Cnidaria, Münster. Guidebook, Excursion A4.

Goldfuss, G.A. 1826. Petrefacta Germaniae. Arnz and Company, Düsseldorf.

Gürich, G. 1909. Leitfossilien, II: Devon. Gebrüder Borntraeger, Berlin, pp. 97–199, Pls. 29–52.

Hall, J., and Whitfield, R.P. 1873. Descriptions of new species of fossils from the Devonian rocks of Iowa. New York State Cabinet of Natural History, Annual Report 23, pp. 223–239, Pls. 9–12.

He, Yuan-xiang. 1978. Rugosa. *In* Atlas of fossils of southeast China, Sichuan Volume. Part 1. From Sinian to Devonian. *Edited by* Chengdu Institute of Geology and Mineral Resources. Geological Publishing House, Beijing, pp. 98–179, 555–568, Pls. 51–88. (In Chinese).

He, Yuan-xiang, and Fan, Ying-nian. 1988. Rugosa. *In* Devonian stratigraphy, paleontology and sedimentary facies of Longmenshan, Sichuan. *Edited by* Hou, Hong-fei et al. Geological Publishing House, Beijing, pp. 165–195, Pls. 17–37. (In Chinese).

Hill, D. 1939. The Devonian rugose corals of Lilydale and Loyola, Victoria. Proceedings of the Royal Society of Victoria, **51**: 219–256, Pls. 13–16.

Hill, D. 1941. The Lower Middle Devonian rugose corals of the Murrumbidgee and Goodradigbee Rivers, N.S.W. Journal and Proceedings of the Royal Society of New South Wales, **74**: 247–276, Pls. 9–11.

Hill, D. 1942*a*. Middle Palaeozoic rugose corals from the Wellington district, N.S.W. Journal and Proceedings of the Royal Society of New South Wales, **76**: 182–189, Pls. 5, 6.

Hill, D. 1942*b*. The Devonian rugose corals of the Tamworth district, N.S.W. Journal and Proceedings of the Royal Society of New South Wales, **76**: 140–164, Pls. 2–4.

Hill, D. 1954. Coral faunas from the Silurian of New South Wales and the Devonian of Western Australia. Bureau of Mineral Resources, Geology and Geophysics, Australia, Bulletin 23.

Hill, D. 1956. Rugosa. *In* Treatise on Invertebrate Paleontology. Part F. Coelenterata. *Edited by* R.C. Moore. Geological Society of America and University of Kansas Press, Lawrence, pp. F233–F324.

Hill, D. 1981. Treatise on Invertebrate Paleontology. Part F. Coelenterata. Supplement 1, Rugosa and Tabulata. *Edited by* C. Teichert. Geological Society of America and University of Kansas Press, Boulder and Lawrence, 2 volumes.

Hill, D., and Jell, J.S. 1970. Devonian corals from the Canning Basin, Western Australia. Geological Survey of Western Australia, Bulletin 121.

Hubmann, B. 1992. Die Korallenfauna aus dem Devon von Feke (Antitaurus, SE-Türkei). II, Rugosa. Geologisch-Paläontologische Mitteilungen Innsbruck, **18**: 151–169.

International Commission on Zoological Nomenclature 1999. International Code of Zoological Nomenclature. 4th Edition. International Trust for Zoological Nomenclature, London.

Ivaniya, V.A. 1952. Verkhnedevonskie korally rugosa Glubokinskogo izvestnyaka c. Solomino (Kuzbass). Tomskiy Gosudarstvenniy Universitet, Uchenye Zapiski 18, pp. 125–164, Pls. 1–8.

Ivaniya, V.A. 1965. Devonskie korally rugosa Sayano-Altayskoy gornoy oblasti. Tomskiy Universitet, Tomsk.

Ivaniya, V.A. 1980. Devonskie chetyrekhluchevye korally Salaira i severnogo Kuzbassa (sistematika, biostratigrafiya, ekologiya). Tomskiy Universitet, Tomsk.

Ivanovskiy, A.B., and Shurygina, M.V. 1980. Reviziya devonskikh rugoz Urala. Paleontologicheskiy Institut Akademiya Nauk SSSR, Trudy 186.

Jia, Hui-zhen. 1977. Devonian rugose corals. *In* Paleontological atlas of central and southern China, Part 2, Late Paleozoic. *Edited by* Geological Research Institute, Hubei et al. Geological Publishing House, Beijing, pp. 112–168, 238–240, 764–777, Pls. 39–41, 43–60, 63, 65, 71. (In Chinese).

Jin, Shui-gen. 1982. Descriptions of corals. *In* Paleontological atlas of Hunan Province. *Edited by* Hunan Provincial Geological Bureau. Geological Publishing House, Beijing, pp. 81–162, 820–842, Pls. 46–112. (In Chinese).

Jin, Shan-yu, and He, Jin-han. 1982. The Devonian rugose corals of Guangxi, their sequence and systematic descriptions. *In* The Devonian biostratigraphy of Guangxi and adjacent area. *Edited by* Shun-liang Bai, Shan-yu Jin, and Zhong-shan Ning. Peking University Press, Beijing, pp. 109–148, 160–165, Pls. 22–36. (Imprint, 1981. In Chinese).

Joseph, J., and Tsien, H.H. 1975. Calcaires mésodévoniens et leurs faunes de Tétracorallaires en haute Vallée d'Ossau (Pyrénées-Atlantiques). Bulletin de la Société d'Histoire Naturelle de Toulouse, **3**: 179–203, Pls. 1–6.

Khoa, N.D., and Thom, N. 1980. Phu lop San ho bon tia (Rugosa). *In* Hoa thach dac trung o mien bac Viet Nam. Nha xuat ban khoa hoc va ky thuat, Hanoi, pp. 86–94, Pls. 22–30. (In Vietnamese).

Klapper, G. 1975. The Independence Shale problems. *In* Trip #1: Devonian limestone facies: Cedar Valley and State Quarry limestones in the Iowa City region. 39th Annual Tri-State Geological Field Conference, Guidebook, pp. 8–9.

Klapper, G. 1989. The Montagne Noire Frasnian (Upper Devonian) conodont succession. *In* Devonian of the world. *Edited by* N.J. McMillan, A.F. Embry, and D.J. Glass. Canadian Society of Petroleum Geologists, Memoir 14, Volume 3, pp. 449–468. (Imprint 1988).

Klapper, G., and Becker, R.T. 1999. Comparison of Frasnian (Upper Devonian) conodont zonations. Bollettino della Società Paleontologica Italiana, **37**: 339–348.

Klapper, G., and Lane, H.R. 1985. Upper Devonian (Frasnian) conodonts of the *Polygnathus* biofacies, N.W.T., Canada. Journal of Paleontology, **59**: 904–951.

Kong, Lei. 1981. Middle Devonian rugose corals in organic reef from Dachang in Nandan District, Guangxi (southern China). *In* Selected Papers, 12th Annual Conference of the Paleontological Society of China. Science Press, Beijing, pp. 50–60, Pls. 1–3.

Kong, Lei, and Huang, Yun-ming. 1978. Tetracoralla. *In* Paleontological atlas of southwestern China, part of Sichuan Province and part of Guizhou Province. *Edited by* Guizhou Stratigraphy and Paleontology Work Team. Geological Press, Beijing, pp. 35–161, 739–759, Pls. 13–55. (In Chinese).

Lakhov, G.V. 1981. Novye vidy kolonialnykh devonskikh rugoz Novoy Zemli. Leningradskiy Gorniy Institut, Zapiski 85, pp. 65–74.

Lakhov, G.V. 1986. Biostratigraficheskoe raschlenenie Verkhnedevonskikh (Franskikh) otlozheniy Novoy Zemli po rugozam. *In* Stratigrafiya i fauna Novoy Zemli. Sbornik nauchnykh trudov. *Edited by* V.I. Bondarev. Sevmorgeologiya, Leningrad, pp. 106–121, Pls. 1, 2. (Imprint 1985).

Lambe, L.M. 1901. A revision of the Canadian Palaeozoic corals; the Madreporaria aperosa and the Madreporaria rugosa. Geological Survey of Canada, Contributions to Canadian Palaeontology, Volume 4, Part 2, pp. 97–197, Pls. 6–18.

Lang, W.D., and Smith, S. 1935. *Cyathophyllum caespitosum* Goldfuss, and other Devonian corals considered in a revision of that species. Quarterly Journal of the Geological Society of London, **91**: 538–590, Pls. 35–37.

Lang, W.D., Smith, S., and Thomas, H.D. 1940. Index of Palaeozoic coral genera. British Museum (Natural History), London.

Liao, Wei-hua and Li, Da-qing. 1991. Species of rugose corals from Xihanshui Group of Qinling Mountains. Acta Palaeontologica Sinica, **30**: 601–615, Pls. 1–3. (In Chinese, English summary).

Luke, K.J. 1978. Corals of the Devonian Guilmette Formation from the Leppy Range near Wendover, Utah-Nevada. Brigham Young University, Geology Studies, **25**: 83–98.

Ma, T.Y.H. 1956. A reinvestigation of climate and the relative positions of continents during the Devonian. Research on the past climate and continental drift, Volume 9. T.Y.H. Ma, Taipei.

McLaren, D.J. 1956. Devonian formations in the Alberta Rocky Mountains between Bow and Athabasca Rivers. Geological Survey of Canada, Bulletin 35. (Imprint 1955).

McLaren D.J. 1959. A revision of the Devonian coral genus *Synaptophyllum* Simpson. Geological Survey of Canada, Bulletin 48, pp. 15–33, Pls. 7–10.

McLaren, D.J., Norris, A.W., and McGregor, D.C. 1962. Illustrations of Devonian fossils. Devonian of western Canada. Geological Survey of Canada, Paper 62–4.

McLaren, D.J., Norris, A.W., and Cumming, L.M. 1970. Devonian faunas. *In* Geology and Economic Minerals of Canada. *Edited by* R.J.W. Douglas. Geological Survey of Canada, Economic Geology Report No. 1, pp. 614–622.

McLean, R.A. 1984. Upper Devonian (Frasnian) rugose corals of the Hay River region, Northwest Territories, Canada. Palaeontographica Americana, **54**: 470–474.

McLean, R.A. 1986. The rugose coral *Pachyphyllum* Edwards and Haime in the Frasnian (Upper Devonian) of western Canada. Geological Survey of Canada, Paper 86–1B, pp. 443–455.

McLean, R.A. 1989. Phillipsastreidae (Rugosa) in the Frasnian of western Canada. Association of Australasian Palaeontologists, Memoir 8, pp. 239–249.

McLean, R.A. 1994*a*. Frasnian rugose corals of western Canada. Part 3A: The massive Phillipsastreidae – *Phillipsastrea*, *Chuanbeiphyllum*. Palaeontographica, Abteilung A, **230**: 39–76, Pls. 1–15.

McLean, R.A. 1994*b*. Frasnian rugose corals of western Canada. Part 3B: The massive Phillipsastreidae – *Pachyphyllum*, *Smithicyathus*, *Frechastraea*. Palaeontographica, Abteilung A, **230**: 77–96, Pls. 1–8.

McLean, R.A., and Klapper, G. 1998. Biostratigraphy of Frasnian (Upper Devonian) strata in western Canada, based on conodonts and rugose corals. Bulletin of Canadian Petroleum Geology, **46**: 515–563.

McLean, R.A., and Pedder, A.E.H. 1984. Frasnian rugose corals of western Canada. Part 1: Chonophyllidae and Kyphophyllidae. Palaeontographica, Abteilung A: **184**: 1–38, Pls. 1–13.

McLean, R.A., and Pedder, A.E.H. 1987. Frasnian rugose corals of western Canada. Part 2: The genus *Smithiphyllum*. Palaeontographica, Abteilung A: **195**: 131–173, Pls. 1–15.

McLean, R.A., and Sorauf, J.E. 1989. The distribution of rugose corals in Frasnian outcrop sequences of North America. *In* Devonian of the world. *Edited by* N.J. McMillan, A.F. Embry, and D.J. Glass. Canadian Society of Petroleum Geologists, Memoir 14, Volume 3, pp. 379–396. (Imprint 1988).

McLean, R.A., Marchant, T.R., Jones, B., and Carter, K.M. 1987. The Middle and Upper Devonian of the Hay River region, southern District of Mackenzie, Northwest Territories. Second International Symposium on the Devonian System, Calgary, Alberta. Guidebook, Field Excursion A2.

McLean, R.A., Marchant, T.R., and Bellow, J.M. 1995. The Upper Devonian of the Hay River region, southern District of Mackenzie, N.W.T. First Joint Symposium, Canadian Society of Petroleum Geologists and Canadian Well Logging Society, Calgary, Alberta. Guidebook, Post-Meeting Field Trip 3.

Moore, P.F. 1993. Devonian. Subchapter 4D. *In* Sedimentary cover of the craton of Canada. *Edited by* D.F. Stott and J.D. Aitken. Geological Survey of Canada, Geology of Canada 5, pp. 150–201.

Mountjoy, E.W. 1965. Stratigraphy of the Devonian Miette reef complex and associated strata, eastern Jasper National Park, Alberta. Geological Survey of Canada, Bulletin 110.

Mountjoy, E.W. 1978. Upper Devonian reef trends and configuration of the western portion of the Alberta Basin. *In* The Fairholme carbonate complex at Hummingbird and Cripple Creek. *Edited by* I.A. McIlreath and P.C. Jackson. Canadian Society of Petroleum Geologists, Calgary, pp. 1–30.

Oliver, W.A., Jr. 1976. Noncystimorph colonial rugose corals of the Onesquethaw and Lower Cazenovia Stages (Lower and Middle Devonian) in New York and adjacent states. United States Geological Survey, Professional Paper 869.

Oliver, W.A., Jr. 1993. The Siphonophrentidae (rugose corals, Devonian) of eastern North America. United States Geological Survey, Bulletin 2024-B. (Imprint 1992).

Pedder, A.E.H. 1964. Two new genera of Devonian tetracorals from Australia. Proceedings of the Linnean Society of New South Wales, **88**: 364–367.

Pedder, A.E.H. 1966. The Devonian tetracoral *Haplothecia* and new Australian phacellophyllids. Proceedings of the Linnean Society of New South Wales, **90**: 182–189.

Pedder, A.E.H. 1998. New and revised Lower Devonian Rugosa from western North America and Tasmania. Journal of Paleontology, **72**: 224–245.

Pedder, A.E.H., Jackson, J.H., and Philip, G.M. 1970. Lower Devonian biostratigraphy in the Wee Jasper region of New South Wales. Journal of Paleontology, **44**: 206–251.

Penecke, K.A. 1894. Das Grazer Devon. Jahrbuch der Kaiserlich-Königlichen Geologischen Reichsanstalt, Jahrgang 1893, **43**: 567–616, Pls. 7–12

Penecke, K.A. 1904. Das Sammelergebnis Dr. Franz Schaffer's aus dem Oberdevon von Hadschin im Antitaurus. Jahrbuch der Kaiserlich-Königlichen Geologischen Reichsanstalt, Jahrgang 1903, **53**: 140–152, Pls. 4–7.

Philip, G.M. 1962. The palaeontology and stratigraphy of the Siluro-Devonian sediments of the Tyers area, Gippsland, Victoria. Proceedings of the Royal Society of Victoria, **75**: 124–246.

Pickett, J.W. 1967. Untersuchungen zur Familie Phillipsastreidae. Senckenbergiana Lethaea, **48**: 1–89.

Price, R.A. 1962. Fernie map-area, east half, Alberta and British Columbia. 82G E ½. Geological Survey of Canada, Paper 61–24.

Price, R.A. 1965. Flathead map-area, British Columbia and Alberta. Geological Survey of Canada, Memoir 336.

Reed, F.R.C. 1922. Devonian fossils from Chitral and the Pamirs. Palaeontologica Indica, New Series, **6**: 1–134, Pls. 1–16.

Roemer, C.F. 1876–1883. Lethaea geognostica, 1 Theil. Lethaea palaeozoica. E. Schweizerbart'sche Verlagshandlung (E. Koch), Stuttgart.

Roemer, F.A. 1855. Beiträge zur geologischen Kenntnis des nordwestlichen Harzgebirges. Dritte Abteilung. Palaeontographica, **5**: i–iv, 1–44, Pls. 1–8.

Rohart, J.-C. 1988. Rugueux Givétiens et Frasniens de Ferques (Boulonnais-France). *In* Le Dévonien de Ferques, Bas-Boulonnais (N. France). *Edited by* D. Brice. Biostratigraphie du Paléozoïque, **7**: 231–296, Pls. 28–36.

Rohart, J.-C. 1999. Palaeozoic rugose corals from central and eastern Iran (A.F. de Lapparent and M. Zahedi collections). Annales de la Société Géologique du Nord, **7**: 47–70, Pls. 6–10.

Różkowska, M. 1953. Pachyphyllinae et *Phillipsastraea* du Frasnien de Pologne. Palaeontologica Polonica, **5**: 1–89, Pls. 1–8.

Różkowska, M. 1956. Pachyphyllinae from the Middle Devonian of the Holy Cross Mts. Part 1. Acta Palaeontologica Polonica, **1**: 271–330.

Różkowska, M. 1957. Considerations on Middle and Upper Devonian Thamnophyllidae Soshkina in Poland. Part II. Acta Palaeontologica Polonica, **2**: 81–153.

Różkowska, M. 1960. Blastogeny and individual variation in tetracoral colonies from the Devonian of Poland. Acta Palaeontologica Polonica, **5**: 1–84.

Różkowska, M. 1980. Contribution to the Frasnian tetracorals from Poland. Palaeontologica Polonica, **40**: 3–56, Pls. 1–10. (Imprint 1979).

Sayutina, T.A. 1965. Kolonialnye devonskie rugozy Zakavkazya. Paleontologicheskiy Zhurnal 1965, **4**: 4–10, Pl. 1.

Schouppé, A. 1949. Die "Thamnophyllen" und ihre Bezeihung zur Gruppe des "*Cyathphyllum caespitosum*". Palaeontographica, Abteilung A, **97**: 99–180, Pls. 1–6.

Schouppé, A. 1958. Revision des Formenkreises um *Phillipsastrea* d'Orb., "*Pachyphyllum*" E. & H., *Macgeea* Webst., "*Thamnophyllum*" Pen., *Peneckiella* Soshk. und verwandter Formen. Neues Jahrbuch für Geologie und Paläontologie, Abhandlungen, **106**: 139–244, Pls. 5, 6.

Schouppé, A. 1965. Die Mittel- bis Oberdevonische Korallenfauna von Kuragh (Chitral). Italian Expeditions to the Karakorum (K^2) and Hindu Kush, Scientific Reports 4, pp. 13–53, Pls. 1–3.

Schouppé, A., and Cheng, Y.M. 1969. *Pterorrhiza* Ehrenberg 1834 (Anthozoa Rugosa): proposed suppression under the plenary powers. Z. N. (S.) 1851. Bulletin of Zoological Nomenclature, **25**: 171–172.

Schröder, S. 1996. Revision der rugosen Koralle *Macgeea bathycalyx* (Frech 1886) (Mitteldevon, Rheinisches Schiefergebirge). Senckenbergiana Lethaea, **76**: 53–63.

Scrutton, C.T. 1968. Colonial Phillipsastraeidae from the Devonian of south-east Devon, England. British Museum of Natural History, Bulletin 15, pp. 181–231.

Shurygina, M.V. 1977. Rugozy. *In* Biostratigrafiya i fauna rannego Devona vostochnogo sklona Urala. *Edited by* T.A. Gorokhova. Nedra, Moskva, pp. 43–51, Pls. 11–19.

Smith, S. 1945. Upper Devonian corals of the Mackenzie River region, Canada. Geological Society of America, Special Paper 59.

Song, Xue-liang. 1982. The Carboniferous and Devonian tetracorals from the Baoshan-Shidian region of western Yunnan. *In* Stratigraphy and paleontology of Jinshajiang-Nujiang-Lancanjiang District, 1. Geological Publishing House, Beijing, pp. 18–37, Pls. 1–6. (In Chinese).

Sorauf, J.E. 1972. Middle Devonian coral faunas (Rugosa) from Washington and Oregon. Journal of Paleontology, **46**: 426–439.

Sorauf, J.E. 1987. Upper Devonian (Frasnian) rugose corals from New York State. Journal of Paleontology, **61**: 676–689.

Sorauf, J.E. 1988. Rugose corals from the Frasnian (Upper Devonian) Sly Gap and Contadero Formations of the San Andres Mountains, south-central New Mexico. *In* Contributions to Paleozoic paleontology and stratigraphy. *Edited by* D.L. Wolberg. New Mexico Bureau of Mines and Mineral Resources, Memoir 44, pp. 153–183.

Sorauf, J.E. 1998. Frasnian (Upper Devonian) rugose corals from the Lime Creek and Shellrock Formations of Iowa. Bulletins of American Paleontology, **113**.

Soroka, I.L. 1993. Osobennosti blastogeneza nekotorykh franskikh rugoz. Paleontologicheskiy Zhurnal, 1993, **4**: 15–25.

Soshkina, E.D. 1939. Verkhnedevonskie korally rugosa Urala. Paleontologicheskiy Institut Akademiya Nauk SSSR, Trudy 9.

Soshkina, E.D. 1949. Devonskie korally rugosa Urala. Paleontologiskiy Institut Akademiya Nauk SSSR, Trudy 15.

Soshkina, E.D. 1951. Pozdnedevonskie korally rugosa, ikh systematika i evolyutsiya. Paleontologiskiy Institut Akademiya Nauk SSSR, Trudy 34.

Soshkina, E.D. 1952. Opredelitel devonskikh chetyrekhluchevykh korallov. Paleontologicheskiy Institut Akademiya Nauk SSSR, Trudy 39.

Soshkina, E.D. 1954. Devonskie chetyrekhluchevye korally Russkoy Platformy. Paleontologicheskiy Institut Akademiya Nauk SSSR, Trudy 52.

Spasskiy, N. Ya. 1955. Korally rugosa i ikh znachenie dlya stratigrafii srednego Devona zapadnogo sklona Urala. Vsesoyuznyy Neftyanyy Nauchno-Issledovatelskiy Geologo-Razvedochnyy Institut, Novaya Seriya, Trudy 90, pp. 91–169, Pls. 1–27.

Spasskiy, N. Ya. 1960. Devonskie chetyrekhluchevye korally Rudnogo Altaya. *In* Paleontologicheskoe obosnovanie stratigrafii paleozoya Rudnogo Altaya, 3. *Edited by* D.V. Nalivkin. Gosgeoltekhizdat, Moskva.

Spasskiy, N. Ya. 1977. Devonskie rugozy SSSR (sistematika, stratigraficheskoe i geograficheskoe znachenie). Leningradskiy Universitet, Leningrad.

Stainbrook, M.A. 1946. Corals of the Independence Shale of Iowa. Journal of Paleontology, **20**: 401–427, Pls. 57–61.

Strusz, D.L. 1965. Disphyllidae and Phacellophyllidae from the Devonian Garra Formation of New South Wales. Palaeontology, **8**: 518–571.

Stumm, E.C. 1940. Upper Devonian rugose corals of the Nevada Limestone. Journal of Paleontology, **14**: 57–67, Pls. 7, 8.

Stumm, E.C. 1960. New rugose corals from the Middle and Upper Devonian of New York. Journal of Paleontology, **34**: 161–163, Pl. 30.

Sun, Y.C. 1958. The Upper Devonian coral faunas of Hunan. Palaeontologica Sinica, New Series, B, **8**: 1–28, Pls. 1–12.

Sytova, V.A., and Ulitina, L.M. 1970. Nekotorye pozdneeyfelskie rugozy Zakavkazya. *In* Novye vidy paleozoyskikh mshanok i korallov. *Edited by* G.G. Astrova and I.I. Chudinova. Nauka, Moskva, pp. 117–120, Pl. 43.

Sytova, V.A., and Ulitina, L.M. 1974. Devonskaya sistema. Tetracoralla. *In* Atlas iskopaemoy fauny Armyanskoy SSR. *Edited by* V.T. Akopyana. Akademiya Nauk Armyanskoy SSR, Erevan, pp. 31–39, Pls. 1–6.

Talent, J.A. 1963. The Devonian of the Mitchell and Wentworth Rivers. Geological Survey of Victoria, Memoir 24.

Taylor, P.W. 1951. The Plymouth Limestone. Transactions of the Royal Geological Society of Cornwall, **18**: 146–214.

Thanh, Tong-Dzuy, and Khoa, N.D. 1988. Rugosa. *In* Stratigrafiya i tselenteraty devona Vetnama. 2. Tselenteraty. *Edited by* V.N. Dubatolov. Nauka Sibirskoe Otdelenie, Novosibirsk, pp. 122–144, Pls. 51–60.

Tsien, H.H. 1968. Contribution à l'étude des Disphyllidae (Rugosa) du Dévonien moyen et du Frasnien de la Belgique. Annales de la Société Géologique de Belgique, **91**: 445–474.

Tsien, H.H. 1969. Contribution à l'étude des Rugosa du Couvinien dans la région de Couvin. Institut Géologique de l' Université de Louvain, Mémoire 25.

Tsien, H.H. 1970. Espèces du genre *Disphyllum* (Rugosa) dans le Dévonien Moyen et le Frasnien de la Belgique. Annales de la Société Géologique de Belgique, **93**: 159–182.

Tsyganko, V.S. 1981. Devonskie rugozy Severa Urala. Nauka Leningradskoe Otdelenie, Leningrad.

Vaganova, T.I. 1959. Rugosa. *In* Brakhiopody i korally iz eyfelskikh boksitonosnykh otlozheniy vostochnogo sklona srednego i severnogo Urala. Gosgeoltekhizdat, Moskva, pp. 77–86, Pls. 25–28.

Vinassa de Regny, P. 1919. Coralli mesodevonici della Carnica. Palaeontographica Italica, **24**: 59–120, Pls. 6–12. (Imprint 1918).

Walther, C. 1929. Untersuchungen über die Mitteldevon-Oberdevongrenze. Zeitschrifte der Deutschen geologischen Gesellschafte, **80**: 97–152. (Imprint 1928).

Wang, H.C. 1948. The Middle Devonian rugose corals of eastern Yunnan. Geological Institute, National University of Peking, Contribution 33.

Wang, Xun-lian. 1994. The rugose coral fauna from the upper part of the Heyuanzhai Formation in western Yunnan. Journal of the Faculty of Sciences, Hokkaido University, **23**: 343–553.

Warren, P.S. 1928. The Palaeozoics of the Crowsnest Pass, Alberta. Transactions of the Royal Society of Canada, Section 4, **22**: 109–119, Pl. 1.

Warren, P.S., and Stelck, C.R. 1950. Succession of Devonian faunas in western Canada. Transactions of the Royal Society of Canada, Section 4, **44**: 61–78.

Webster, C.L. 1889. Description of a new genus of corals from the Devonian rocks of Iowa. The American Naturalist, **23**: 710–712.

Whiteaves, J.F. 1891. The fossils of the Devonian rocks of the Mackenzie River basin. Geological Survey of Canada, Contributions to Canadian Palaeontology, Volume 1, Part 3, pp. 197–253.

Workum, R.H., and Hedinger, A.S. 1992. Devonian, Frasnian stratigraphy, Rocky Mountain Front Ranges, Crowsnest Pass to Jasper, Alberta. Geological Survey of Canada, Open File 2509.

Wright, A.J. 1981. A new phillipsastraenid tetracoral from the Devonian of New South Wales. Palaeontology, **24**: 589–608.

Wrzołek, T. 1987. *Rozkowskaella*, a new name for Devonian tetracoral *Trigonella* Różkowska, 1980. Acta Palaeontologica Polonica, **31**: 277. (Imprint 1986).

Wu, Wang-shi, Liao, Wei-hua, and Zhao, Jia-ming. 1982. Palaeozoic corals from Xizang. *In* Paleontology of Xizang. Series of the Scientific Expedition to the Qinghai-Xizang Plateau, Volume 4, pp. 107–151, Pls. 1–19. Science Press, Beijing. (In Chinese, English summary).

Yoh, S.S. 1937. Die Korallenfauna des Mitteldevons aus der Provinz Kwangsi, Südchina. Palaeontographica, Abteilung A, **87**: 45–76.

Yoh, Sen-shing, and Bai, Shun-liang. 1978. Devonian stratigraphy of Dale, Xiangzhou area, Kwangsi. *In* Symposium on the Devonian System of South China, 1974. *Edited by* Institute for Geological and Mining Research, Chinese Academy of Geological Sciences. Geological Publishing House, Beijing, pp. 43–62, Pls. 5–7. (In Chinese).

Yu, Chang-min, and Kuang, Guo-dun. 1983. Late Middle Devonian rugose corals from Liujing, Heng Xian, Guangxi and their paleoecological significance. Nanjing Institute of Geology and Palaeontology, Academia Sinica, Bulletin 4, pp. 241–278, Pls. 1–8. (Imprint 1982. In Chinese, English summary).

Yu, Chang-min, and Kuang, Guo-dun. 1986. Tetracorals from the Middle-Upper Devonian boundary beds and Upper Devonian in Liujing, Heng Xian of central Guangxi. Nanjing Institute of Geology and Palaeontology, Academia Sinica, Bulletin 7, pp. 127–164, Pls. 1–4. (Imprint 1984. In Chinese, English summary).

Yu, Chang-min, and Liao, Wei-hua. 1978*a*. Lower Devonian rugose corals from Alengchu of Lijiang, northwestern Yunnan. Acta Palaeontologica Sinica, **17**: 245–266, Pls. 1–3. (In Chinese, English summary).

Yu, Chang-min, and Liao, Wei-hua. 1978*b*. Middle Devonian rugose corals of Longdongshui Member, Houershan Formation from Dushan district, Guizhou. Nanjing Institute of Geology and Palaeontology, Memoir 12, pp. 107–149, Pls. 1–16. (In Chinese, English summary).

Yu, Chang-min, Liao, Wei-hua, and Deng, Zhan-qiu. 1974. Devonian corals. *In* A handbook of the stratigraphy and paleontology in southwest China. *Edited by* Nanjing Institute of Geology and Palaeontology. Science Press, Nanjing, pp. 223–232, Pls. 104–112. (In Chinese).

Zheltonogova, V.A., and Ivaniya, V.A. 1961. Devonskaya Sistema. Podklass Tetracoralla (Rugosa). Tetrakorally. *In* Biostratigrafiya paleozoya Sayano-Altayskoy gornoy oblasti. Tom II, Sredniy paleozoy. *Edited by* L.L. Khalfin. Sibirskiy Nauchno-Issledovatelskiy Institut Geologii, Geofiziki i Mineralnogo Syrya 20, pp. 368–408, Pls. 27–57. (Imprint 1960).

Zhen, Yong-yi. 1995. Late Emsian rugose corals of the Mount Podge area, Burdekin Basin, north Queensland. Alcheringa, **19**: 193–234.

Plates

Stratigraphic, geographic, and collection details for all figured Canadian material are provided in the body of the text with the descriptions of the relevant species, and in the Locality Register. For illustrated non-Canadian specimens, these data are given in the plate explanations.

Plate 1

Figs. 1–10. *Macgeea parva* Webster

1. SUI 100605, Stainbrook Collection 20932, Independence Shale, Brandon #1, Buchanan Co., Iowa. Transverse section, ×5.

2, 3. SUI 100603, Stainbrook Collection 20933, Independence Shale, Brandon #3, Buchanan Co., Iowa. Transverse and longitudinal sections, ×5.

4, 8. GSC 126350. Transverse and longitudinal sections, ×5.

5. GSC 126352. Longitudinal section, ×5.

6. SUI 100604, Stainbrook Collection 20932, Independence Shale, Brandon #1, Buchanan Co., Iowa. Longitudinal section, ×5.

7. GSC 126351. Transverse section, ×5.

9, 10. GSC 126353. Transverse and longitudinal sections, ×5.

Figs. 11–16. *Macgeea proteus* Smith

11, 12. GSC 9300, *holotype.* Transverse and longitudinal sections, ×3.

13–15. GSC 126359. Transverse and longitudinal sections, ×3.

16. GSC 126358. Longitudinal section, ×3.

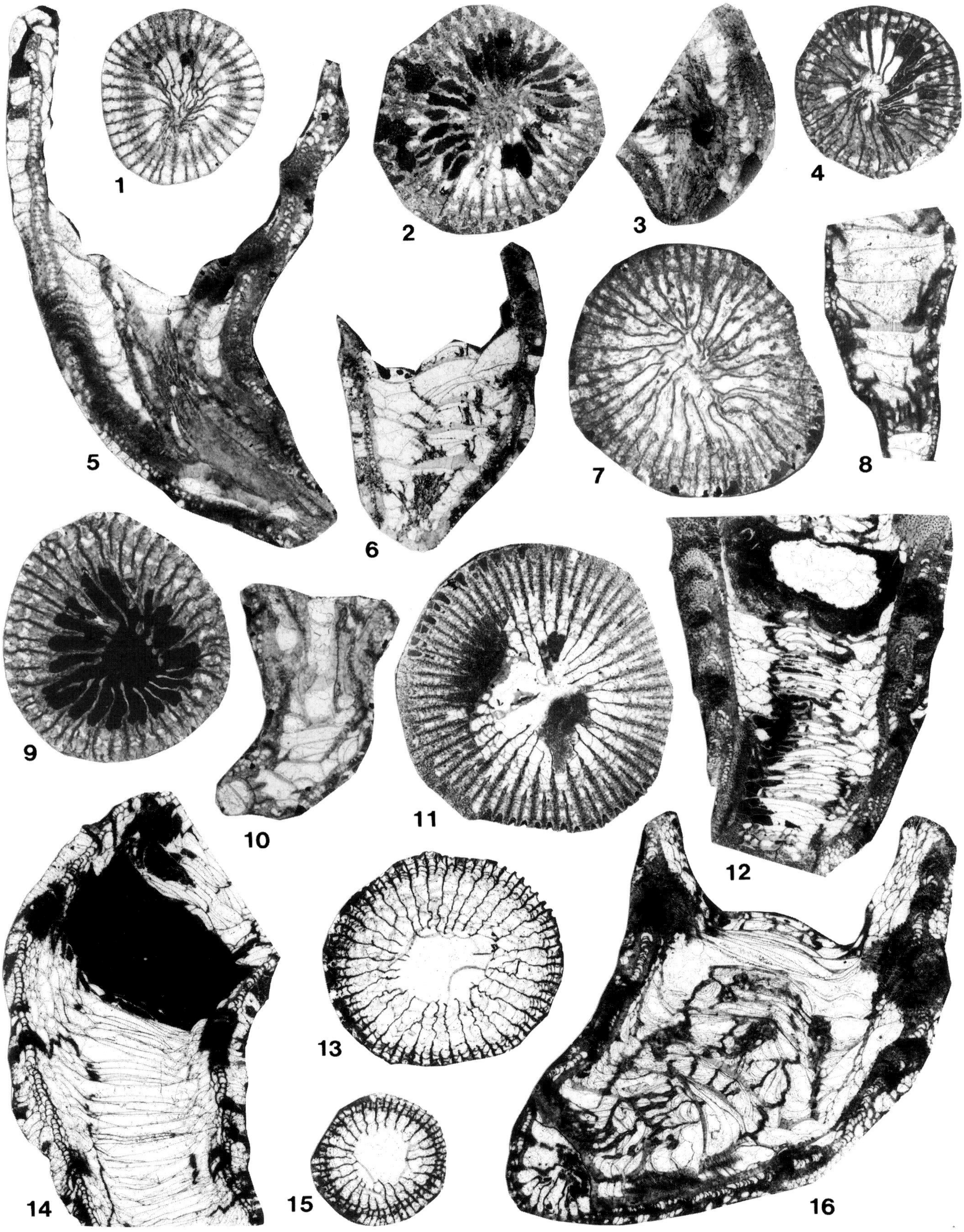
1
2
3
4
5
6
7
8
9
10
11
12
13
14
15
16

Plate 2

Figs. 1–9. *Macgeea proteus* Smith

1, 2. GSC 126356. Transverse and longitudinal sections, ×3. Fig. 2 shows lateral extension of outer dissepimentarium anchoring the corallum to a tabular stromatoporoid.

3, 6, 9. GSC 126355. Transverse and longitudinal sections, ×3.

4, 5, 7. GSC 126354. Figs. 4, 5, longitudinal and transverse sections, ×3; Fig. 7, portion of longitudinal section in Fig. 4 enlarged to show character of rhipidacanthine trabeculae, ×8.

8. GSC 126360. Unsectioned specimen in lateral view showing two calicinal offsets. Note partial preservation of epitheca, ×1.5.

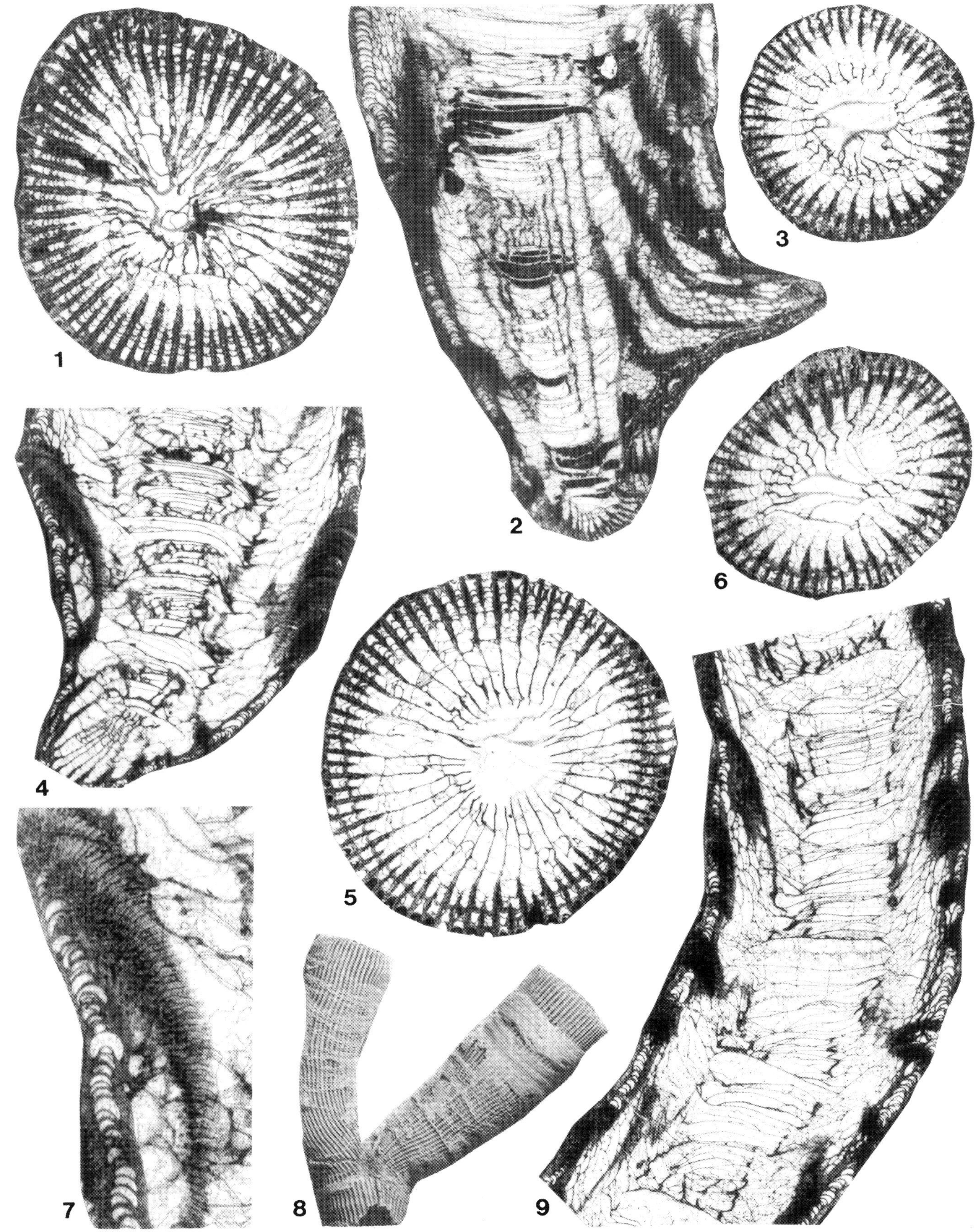
1
2
3
4
5
6
7
8
9

Plate 3

Figs. 1, 4, 6, 9. *Macgeea telopea* Crickmay

1, 4. PRI 27076, *holotype.* Transverse and longitudinal sections, ×3.

6, 9. GSC 126361. Transverse and longitudinal sections, ×3.

Fig. 2. *Macgeea proteus* Smith

GSC 126357. Longitudinal section, ×3.

Figs. 3, 5, 7, 8, 10–12. *Macgeea soraufi* n. sp.

3, 5. GSC 126366, *paratype.* Longitudinal and transverse sections, ×3.

7. GSC 126368, *paratype.* Longitudinal section of juvenile specimen, ×3.

8, 11, 12. GSC 126363, *paratype.* Transverse sections, ×3.

10. GSC 126364, *paratype.* Longitudinal section, ×3.

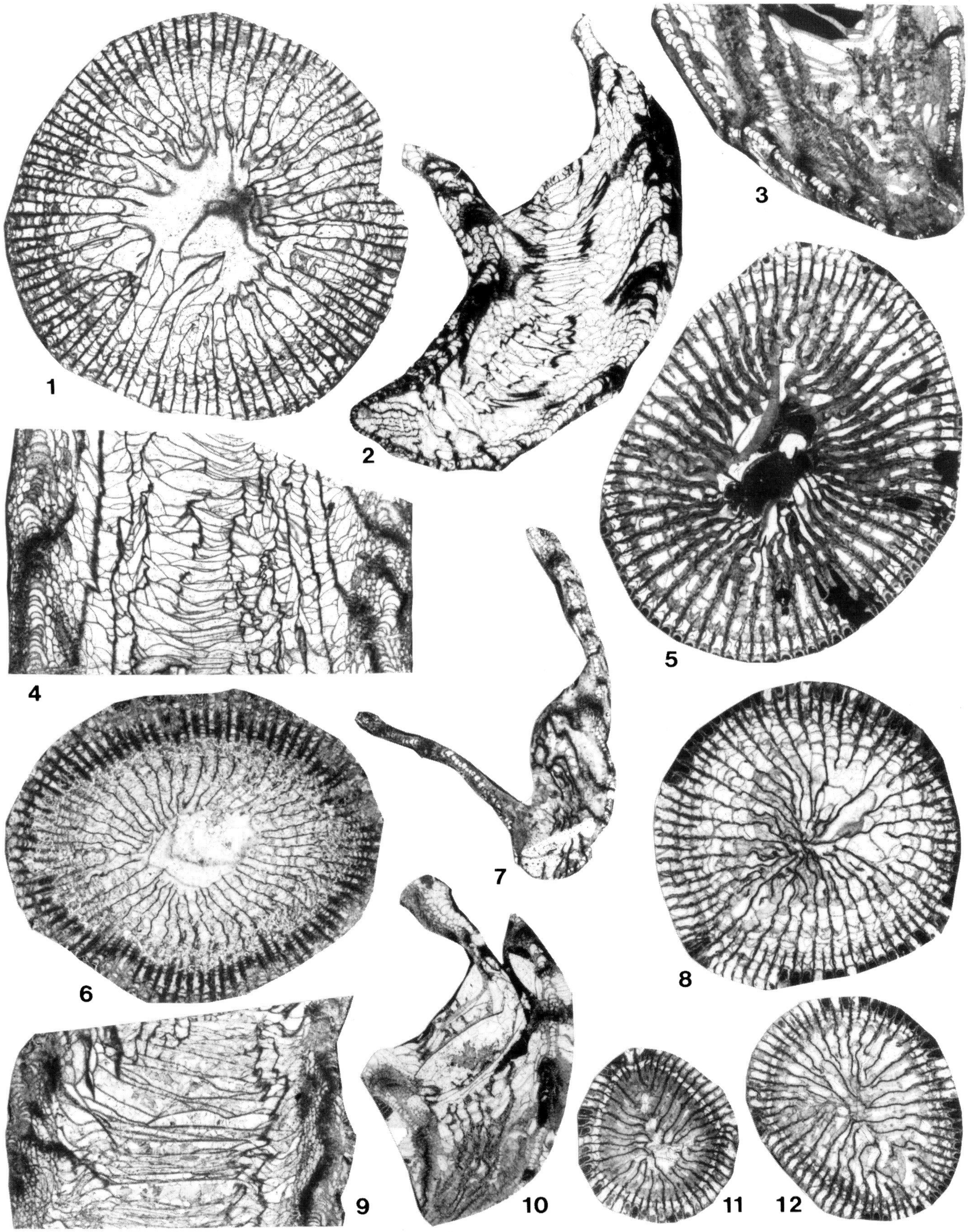
1
2
3
4
5
6
7
8
9
10
11
12

Plate 4

Figs. 1–3, 9. *Macgeea telopea* Crickmay

1, 2. PRI 27078, *paratype*. Transverse and longitudinal sections, ×3.

3, 9. PRI 27077, *paratype*. Transverse and longitudinal sections, ×3.

Figs. 4, 5, 11. *Macgeea soraufi* n. sp.

4, 5. GSC 126367, *paratype*. Transverse and longitudinal sections, ×3.

11. GSC 126365, *paratype*. Longitudinal section, ×3.

Figs. 6–8, 10. *Macgeea pustulosa* n. sp.

GSC 126375, *paratype*. Transverse and longitudinal sections, ×3.

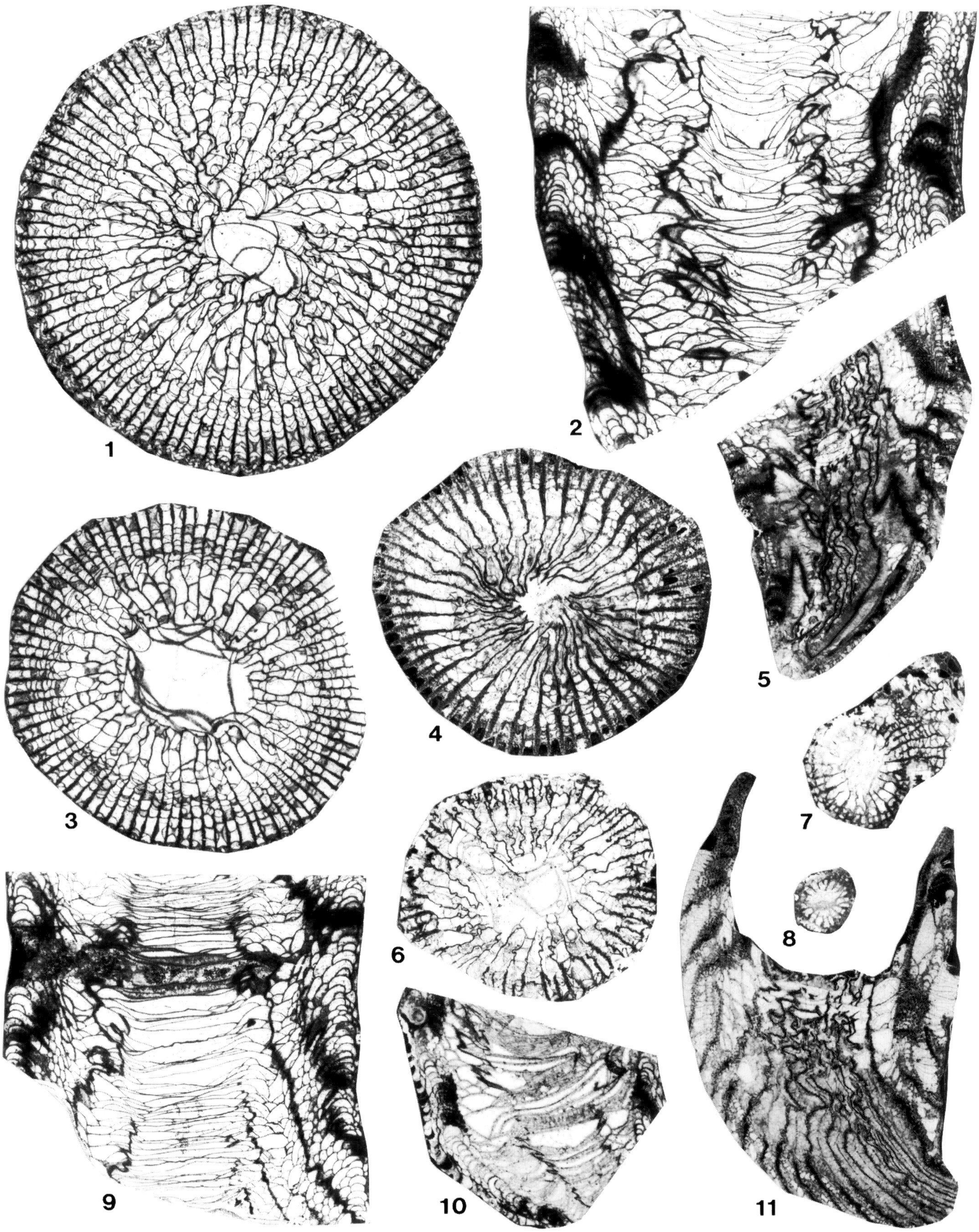
1
2
3
4
5
6
7
8
9
10
11

Plate 5

Figs. 1, 2, 5, 6. *Macgeea sorаufi* n. sp.

1, 5. GSC 126362, *holotype.* Transverse and longitudinal sections, ×3.

2, 6. GSC 126369, *paratype.* Longitudinal and transverse sections, ×3.

Figs. 3, 4, 7–13. *Macgeea pustulosa* n. sp.

3, 4. GSC 126373, *paratype.* Transverse and longitudinal sections, ×3.

7, 9, 10. GSC 9304, *paratype.* Transverse and longitudinal sections, ×3.

8, 11. GSC 126377, *paratype.* Transverse and longitudinal sections, ×3.

12. GSC 126371, *paratype.* Longitudinal section of subpatellate specimen, ×3.

13. GSC 126372, *paratype.* Transverse section, ×3.

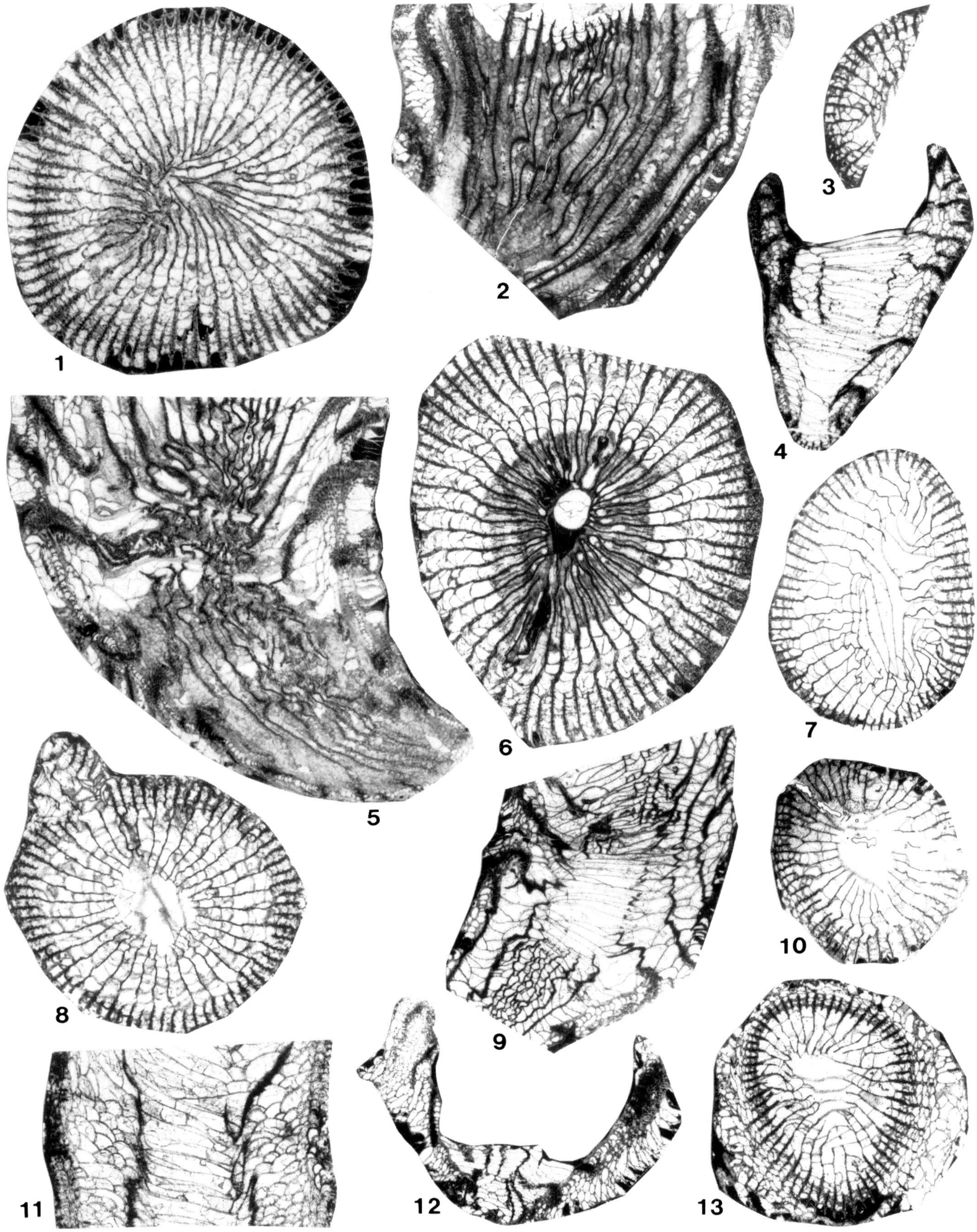
1
2
3
4
5
6
7
8
9
10
11
12
13

Plate 6

Figs. 1, 2. *Macgeea pustulosa* n. sp. ?

GSC 126378. Specimen transitional to *M. telopea* Crickmay. Transverse and abraded longitudinal sections, ×3.

Figs. 3–7. *Macgeea pustulosa* n. sp.

3. GSC 126374, *paratype*. Transverse section, ×3.

4, 5. GSC 126370, *holotype*. Partial transverse and longitudinal sections, ×3.

6, 7. GSC 126376, *paratype*. Transverse and longitudinal sections, ×3.

Figs. 8–15. *Thamnophyllum colemanense* (Warren)

8, 11. GSC 126380. Transverse and longitudinal sections of individual corallites, ×4.

9, 10. GSC 126382. Transverse and longitudinal sections of individual corallites, ×4.

12, 13. UA Dv 647, *lectotype*. Transverse and longitudinal sections of individual corallites, ×4.

14, 15. GSC 126393. Transverse and longitudinal sections, ×4.

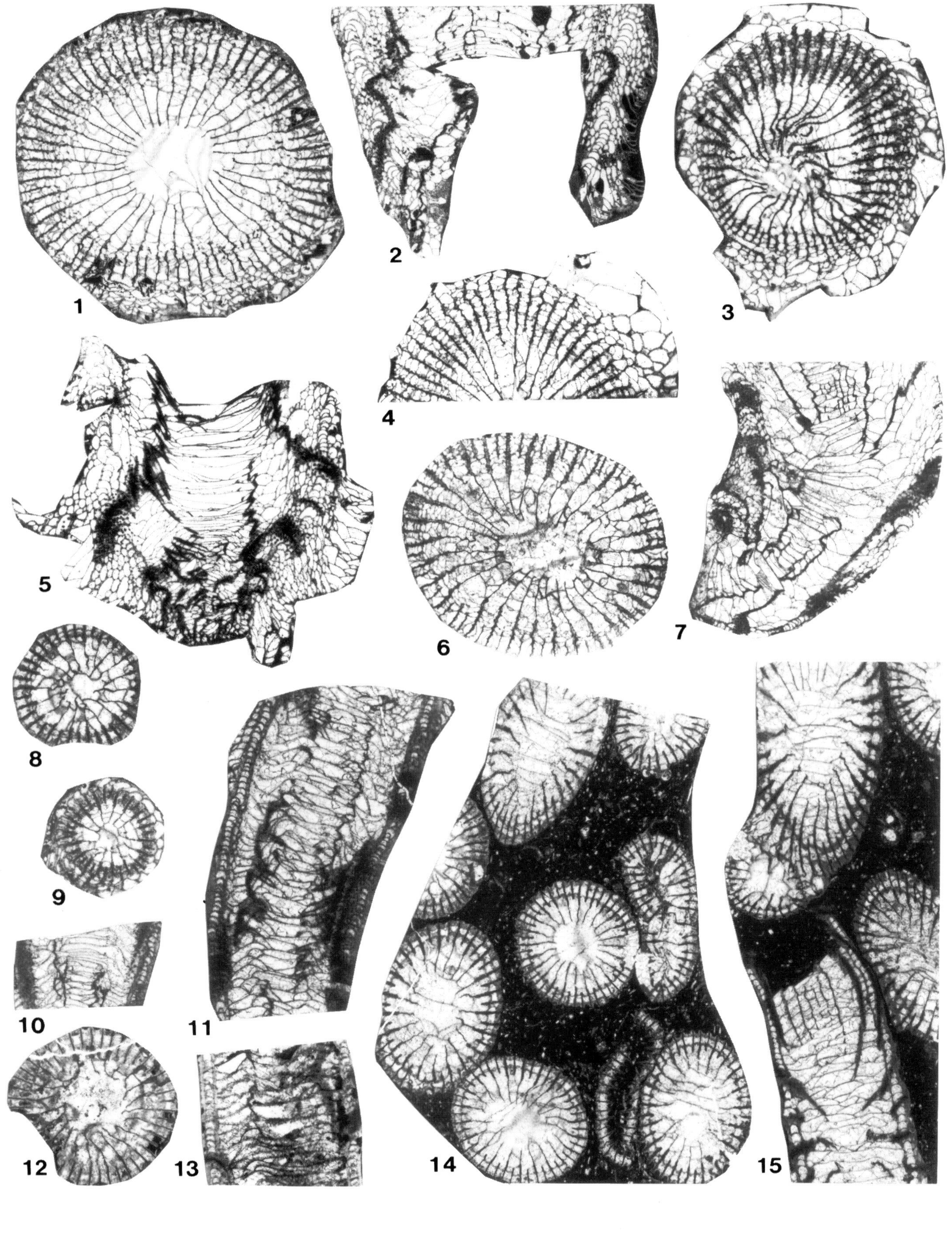
1
2
3
4
5
6
7
8
9
10
11
12
13
14
15

Plate 7

Figs. 1–7. *Thamnophyllum colemanense* (Warren)

1–3. GSC 126387. Transverse and longitudinal sections, ×4.

4, 7. GSC 126391. Longitudinal and transverse sections of specimen with well-preserved outer dissepiments, ×4.

5, 6. GSC 126388, possible topotype. Transverse and longitudinal sections, ×4.

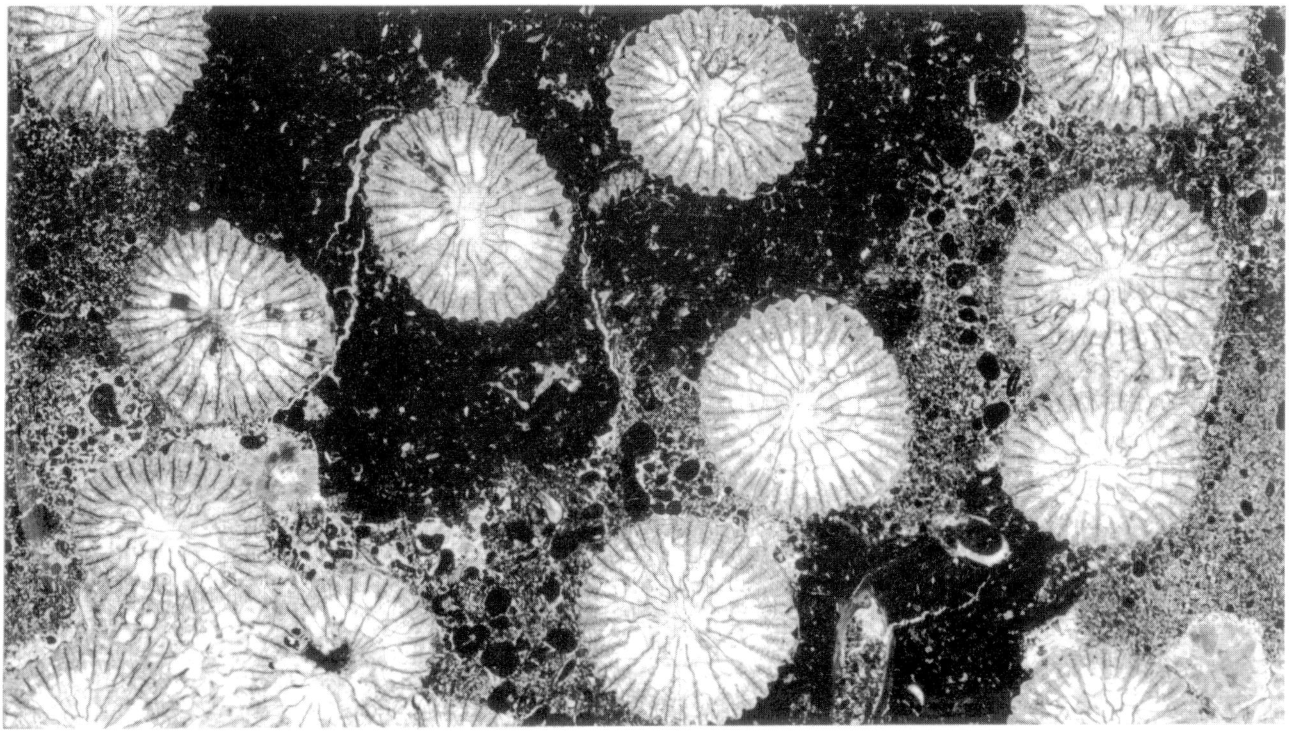

1

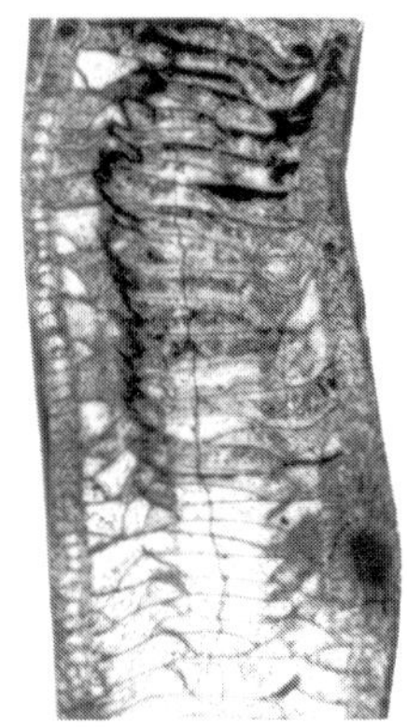

2

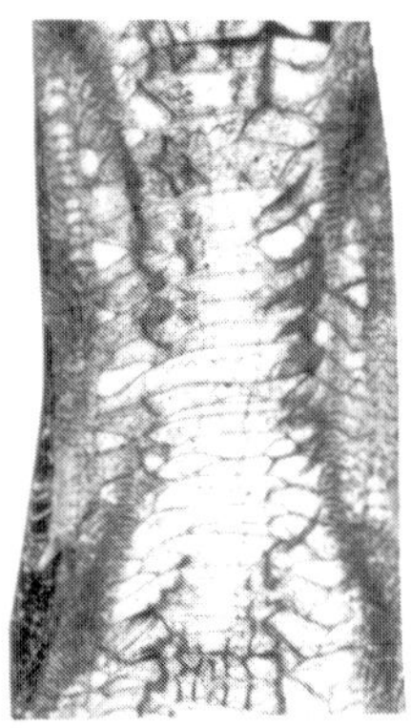

3

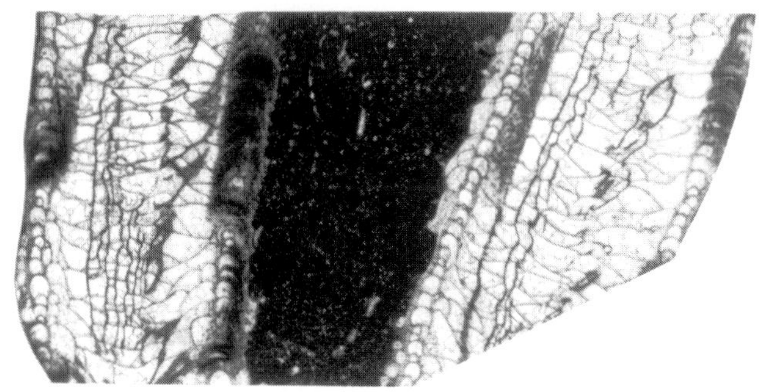

4

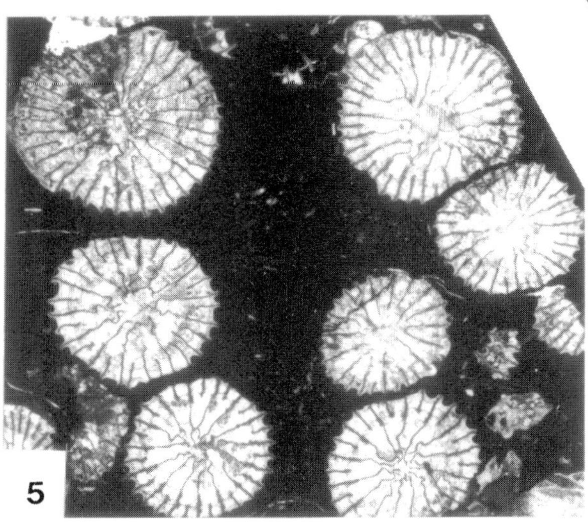

5

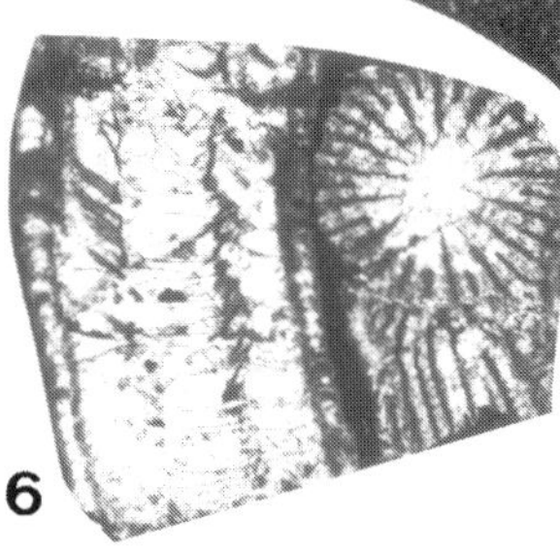

6

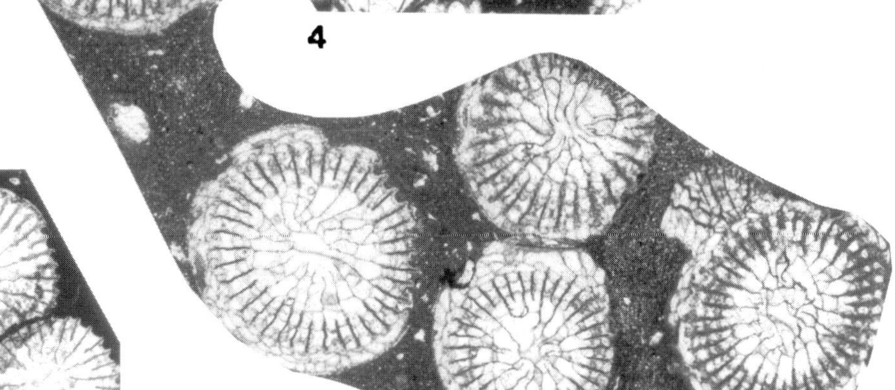

7

Plate 8

Figs. 1–9. *Thamnophyllum colemanense* (Warren)

1, 2. GSC 126389. Transverse and longitudinal sections, ×4.

3–6, 8. GSC 9330a, holotype of *Phacellophyllum fenense* McLaren. Transverse and longitudinal sections, ×4.

7, 9. GSC 126385. Longitudinal and transverse sections of specimen with septa unusually dilated peripherally, ×4.

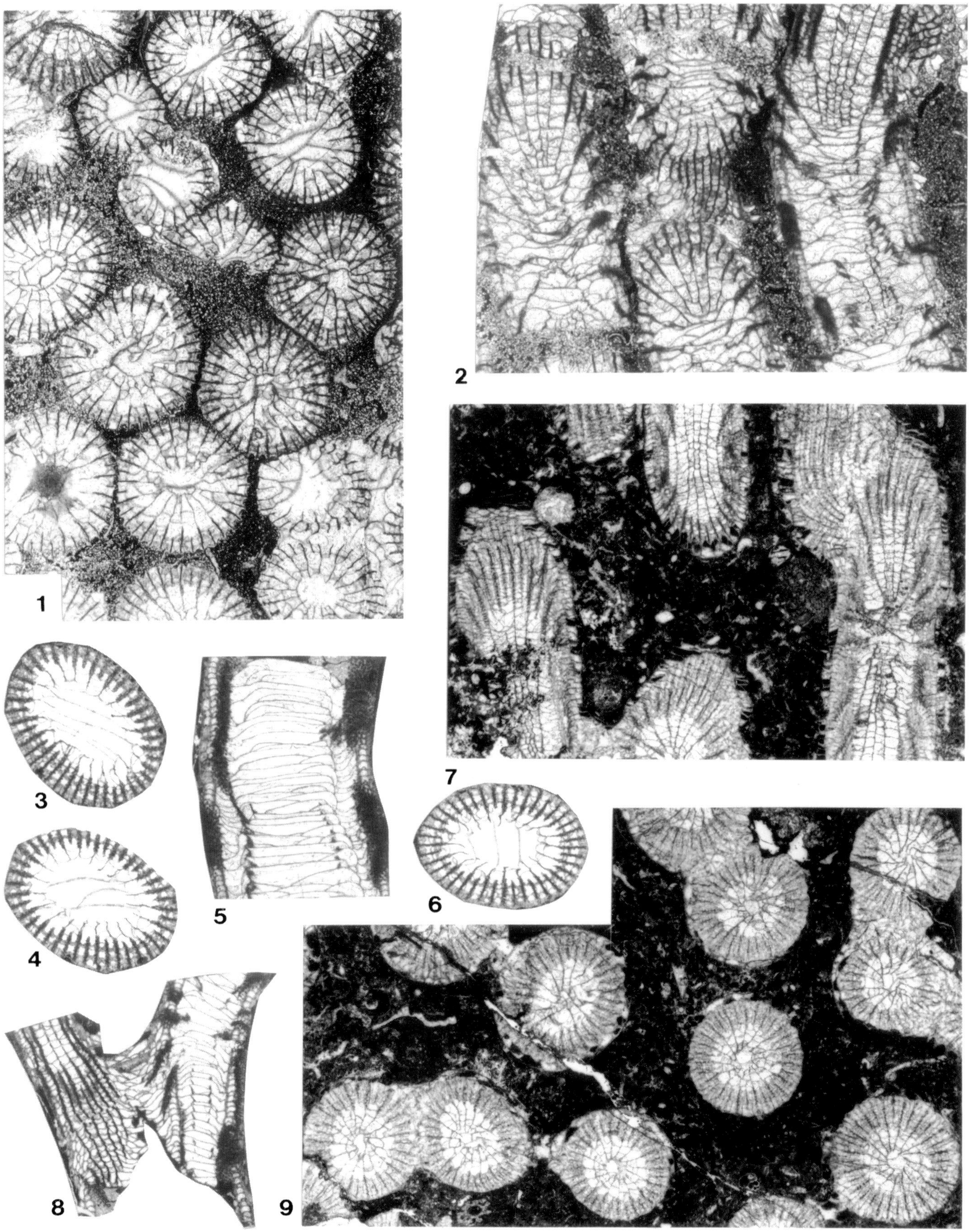
1
2
3
4
5
6
7
8
9

Plate 9

Figs. 1–7, 11. *Thamnophyllum colemanense* (Warren)

1–4. GSC 126379. Transverse and longitudinal sections of specimen with extreme development of outer dissepiments, ×4.

5, 7. GSC 126390. Transverse and longitudinal sections, ×4.

6, 11. GSC 126392. Longitudinal and transverse sections, ×4.

Figs. 8–10, 12, 13. *Thamnophyllum tructense* (McLaren)

GSC 126398. Longitudinal and transverse sections, ×6.

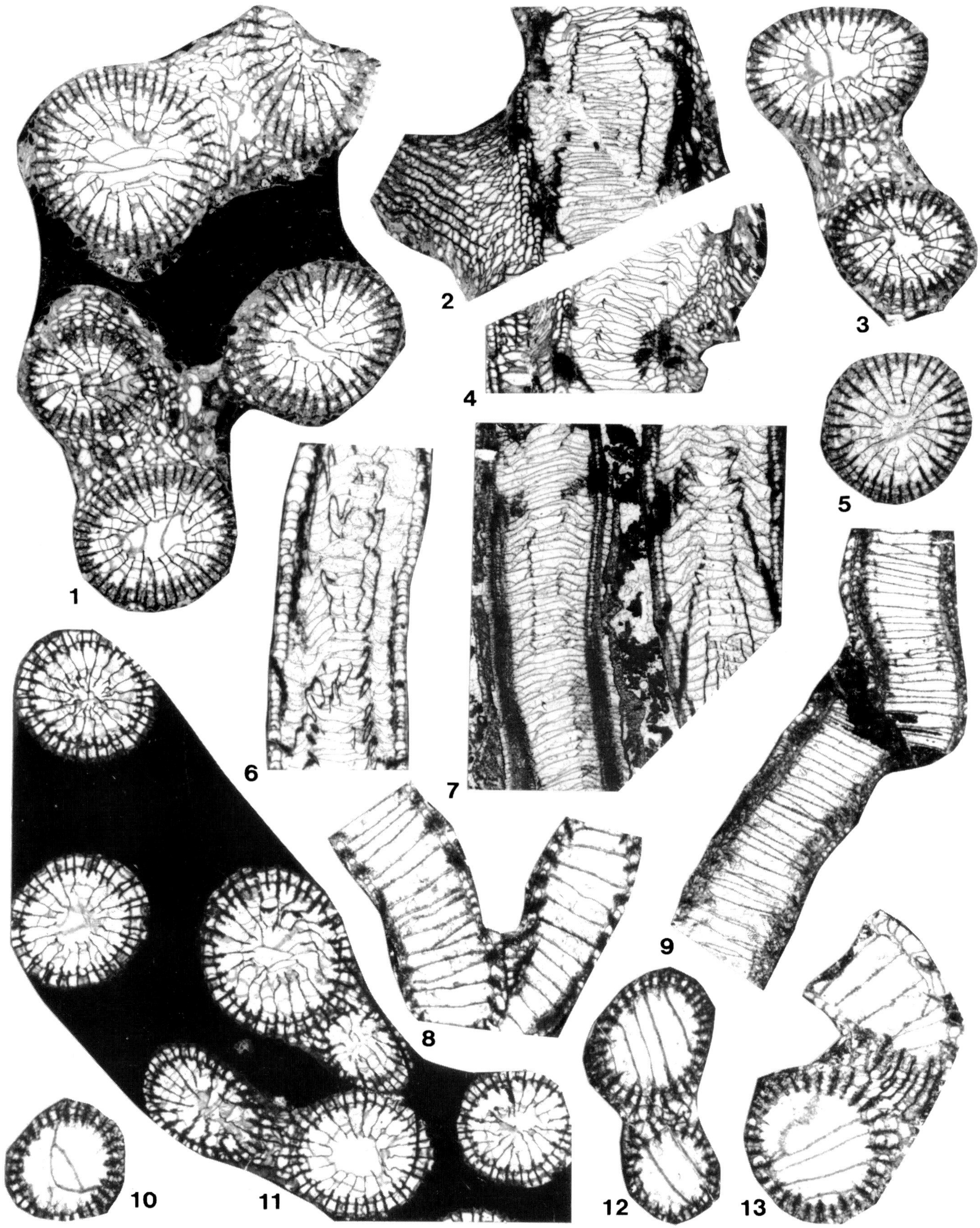
1
2
3
4
5
6
7
8
9
10
11
12
13

Plate 10

Figs. 1–10. *Thamnophyllum tructense* (McLaren)

1, 2. GSC 126399. Transverse and longitudinal sections, ×6.

3–6. GSC 6310, *holotype*. Transverse and longitudinal sections of individual corallites, ×6.

7, 9. GSC 126397. Longitudinal and transverse sections, ×6.

8, 10. GSC 126394. Longitudinal and transverse sections, ×6.

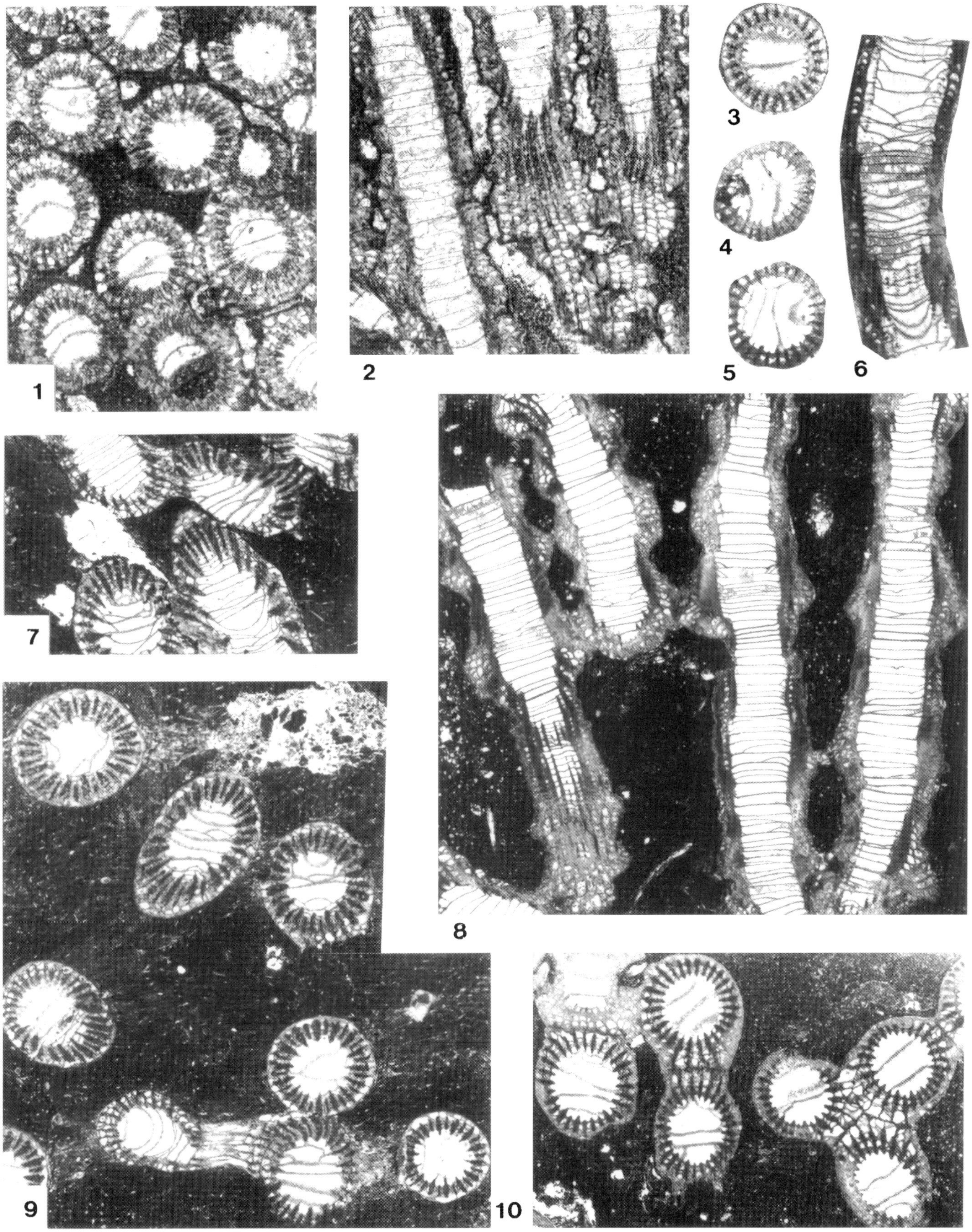
1
2
3
4
5
6
7
8
9
10

Plate 11

Figs. 1–4. *Thamnophyllum tructense* (McLaren)

1, 2. GSC 126395. Longitudinal and transverse sections of specimen with abundant outer dissepiments connecting corallites, ×6.

3, 4. GSC 126396. Transverse and longitudinal sections, ×6.

Figs. 5–9. *Thamnophyllum pedderi* n. sp.

5–7. GSC 126403, *paratype*. Transverse and longitudinal sections, ×6.

8, 9. GSC 16722, *paratype*. Transverse and longitudinal sections, ×6.

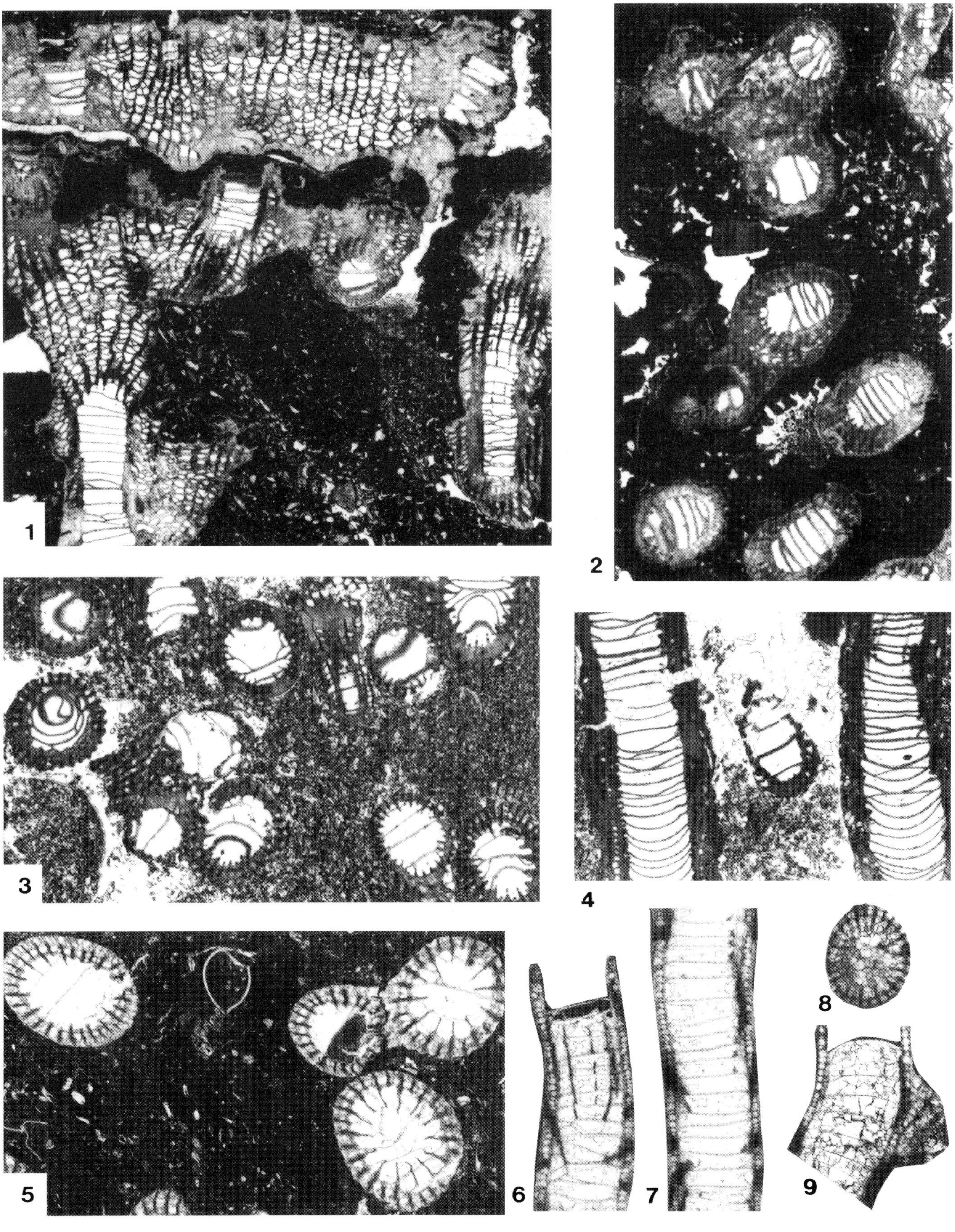

Plate 12

Figs. 1–8. *Thamnophyllum pedderi* n. sp.

1–3. GSC 126400, *holotype*. Transverse and longitudinal sections, ×6.

4–6. GSC 126404, *paratype*. Longitudinal and transverse sections, ×6.

7, 8. GSC 126401, *paratype*. Transverse and longitudinal sections of specimen with unusually small corallites, short major septa, and peripheral septal dilation and stereome, ×6.

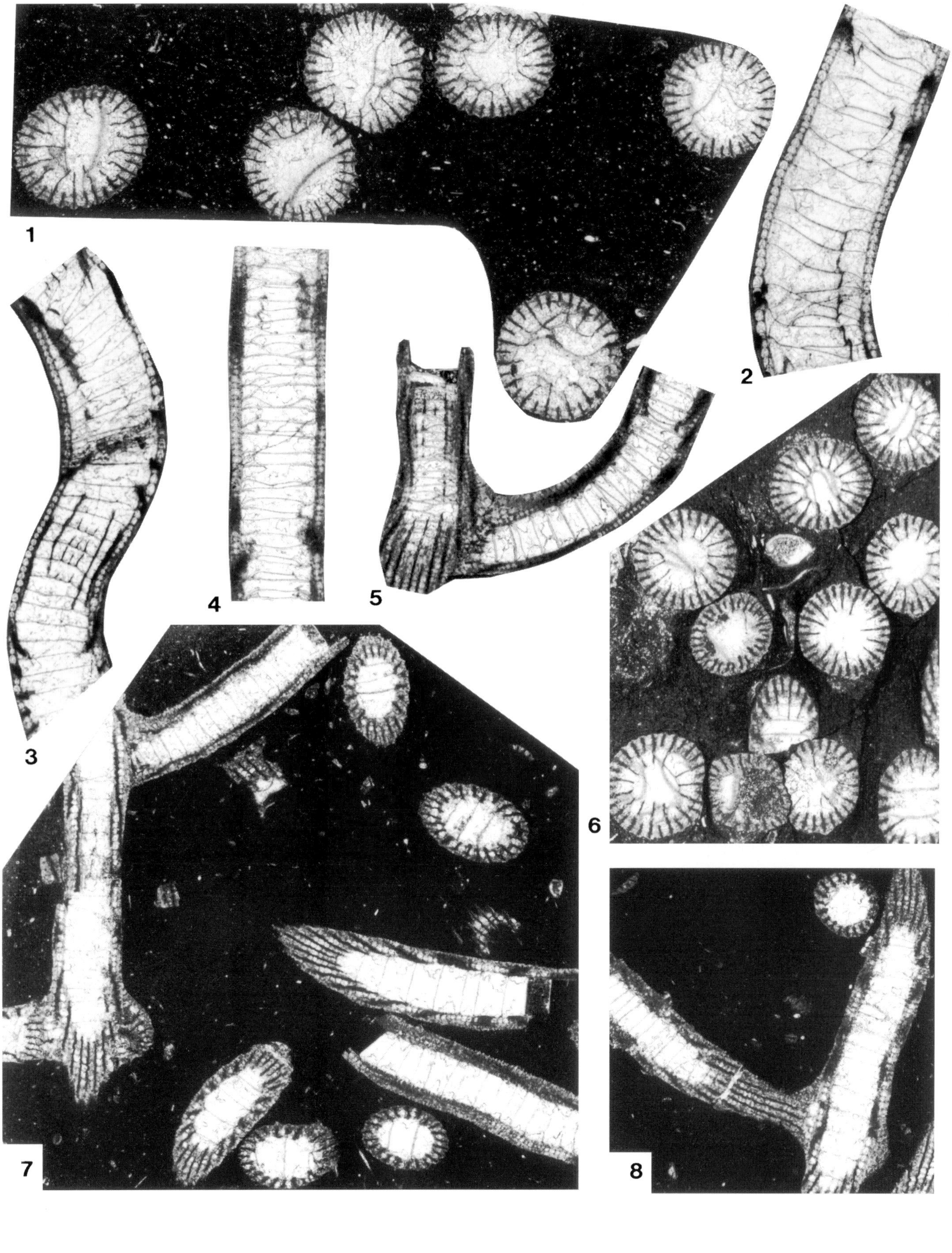

Plate 13

Figs. 1–6. *Thamnophyllum pedderi* n. sp.

GSC 126402, *paratype*. Transverse and longitudinal sections, ×6.

Figs. 7, 8. *Thamnophyllum cordense* n. sp.

GSC 126409, *holotype*. Transverse sections, ×3. See also Pl. 14, Figs. 2, 4.

1
2
3
4
5
6
7
8

Plate 14

Figs. 1, 3, 5–8. *Thamnophyllum julli* n. sp.

1, 3, 5, 6. GSC 126413, *paratype*. Transverse and longitudinal sections, ×3.

7, 8. GSC 126411, *paratype*. Transverse and longitudinal sections, ×3.

Figs. 2, 4. *Thamnophyllum cordense* n. sp.

GSC 126409, *holotype*. Longitudinal sections, ×3. See also Pl. 13, Figs. 7, 8.

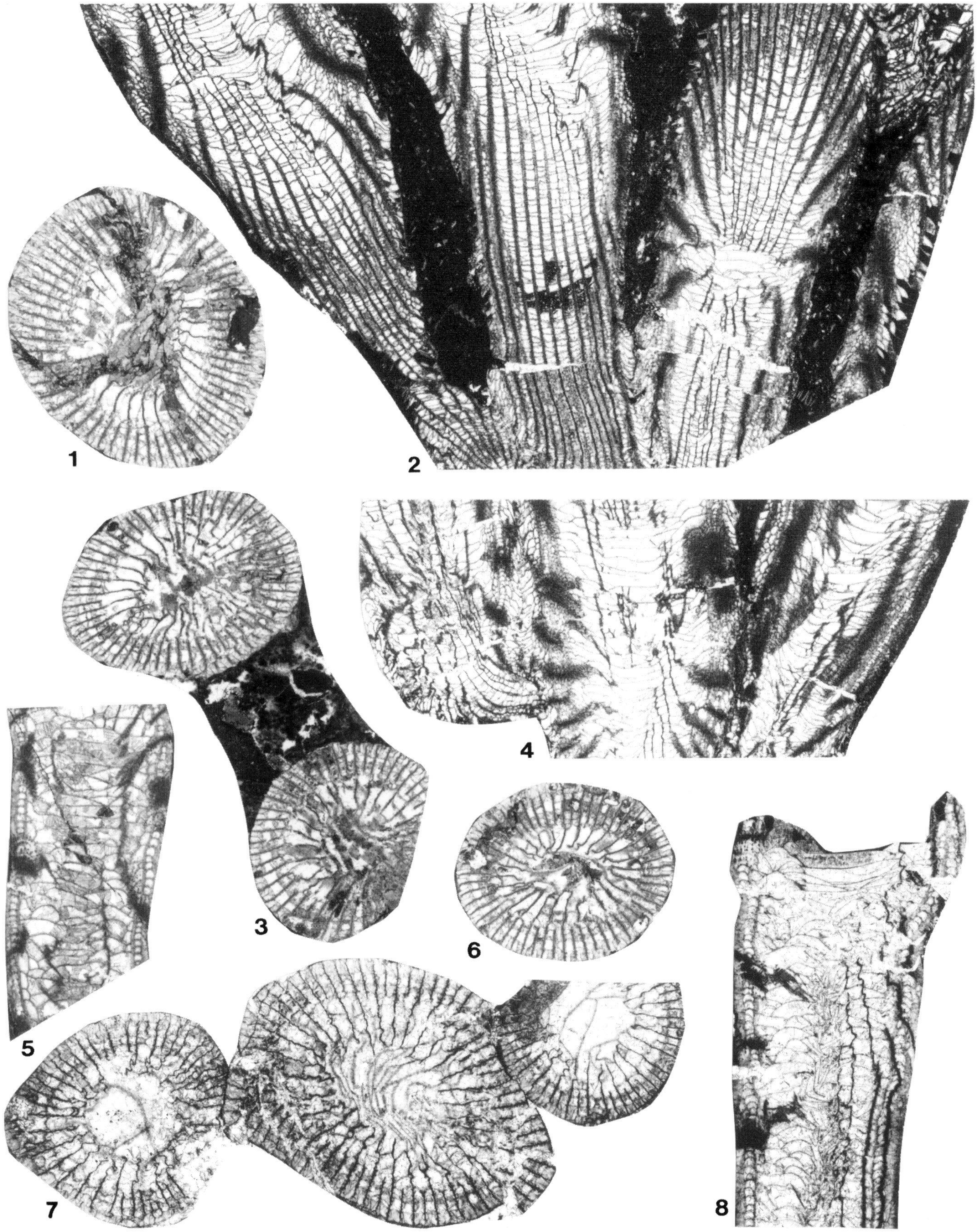
1
2
3
4
5
6
7
8

Plate 15

Figs. 1–6, 8, 9. *Thamnophyllum julli* n. sp.

1, 4, 5. GSC 126414, *paratype*. Transverse and longitudinal sections, ×3.

2, 3. GSC 126412, *paratype*. Transverse and longitudinal sections of individual corallites, ×3.

6, 8, 9. GSC 126410, *holotype*. Longitudinal and transverse sections, x3.

Fig. 7. *Peneckiella minor* (Roemer)

GSC 126417, approximate topotype. Fragmented colony showing transverse, oblique and longitudinal sections, ×4. Note the combination of peneckielloid, horseshoe, and globose dissepiments in the longitudinal section. Ibergerkalk, mid Frasnian, Winterberg Quarry (Stop 1, Gischler *in* Gischler et al. 1991, p. 25), Harz Mountains, Germany. Collected by the writer, 1991.

1

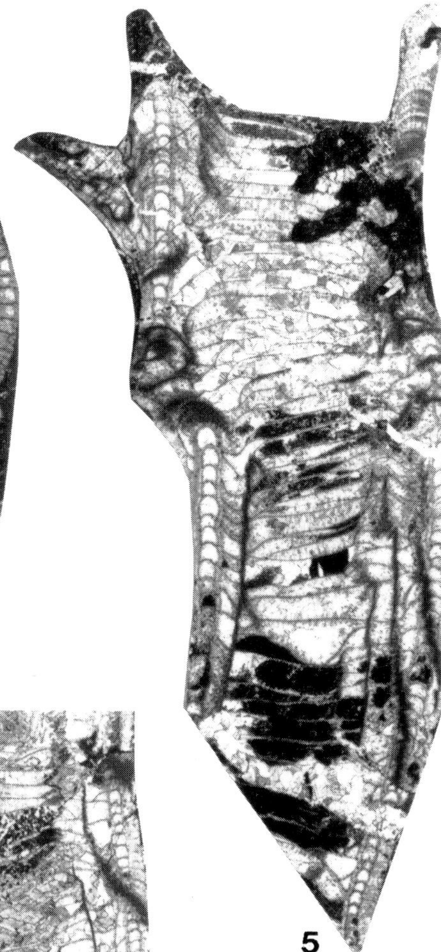

5

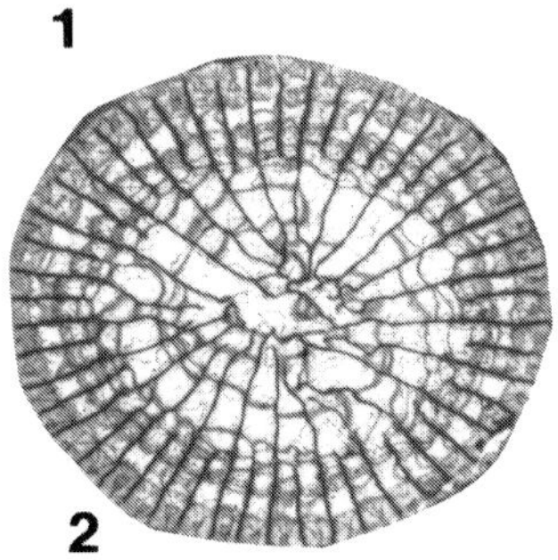

2

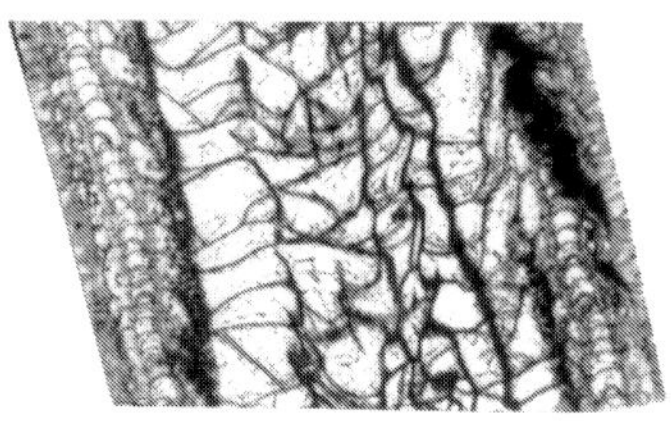

3

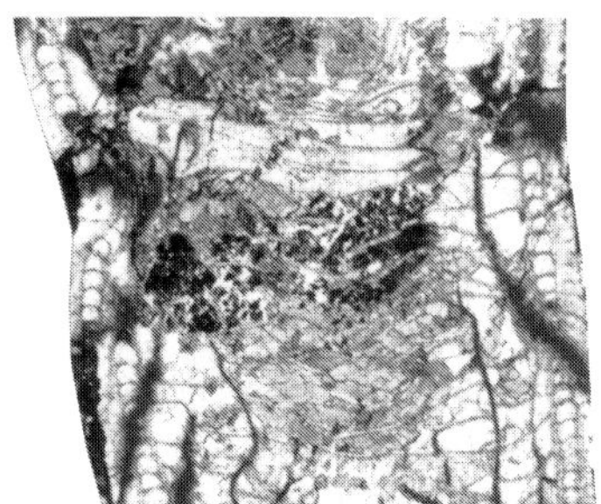

4

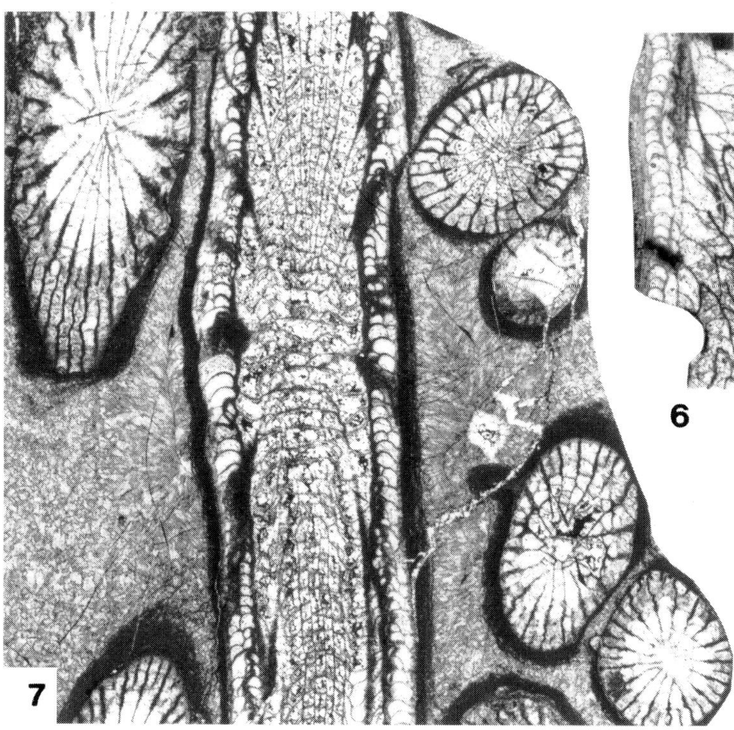

7

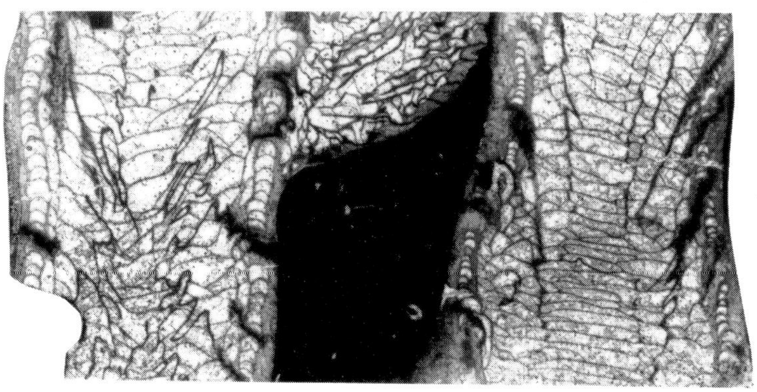

6

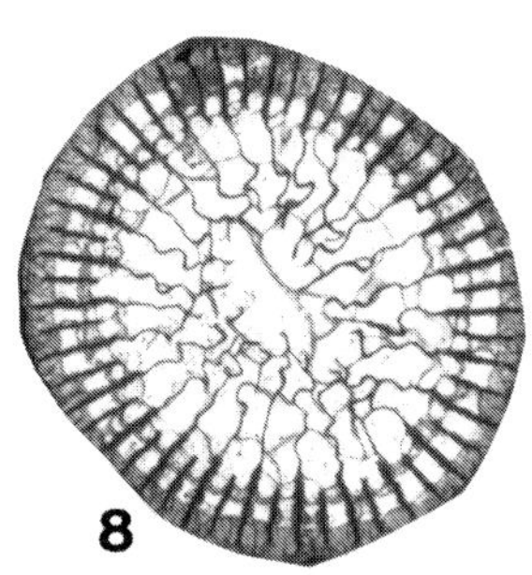

8

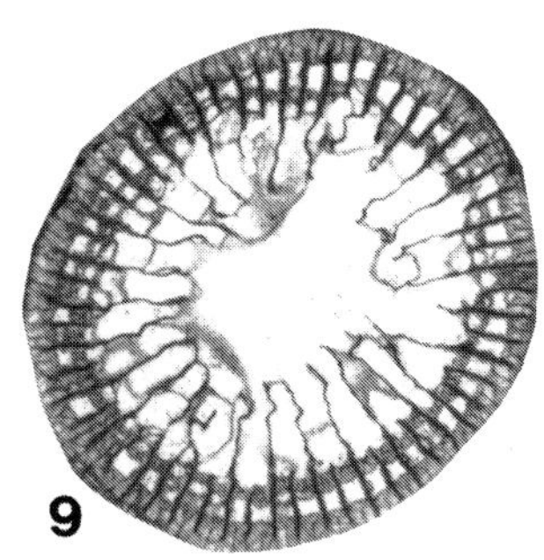

9

Plate 16

Figs. 1–6. *Peneckiella minor* (Roemer)

1–3. GSC 126416, approximate topotype. Transverse and longitudinal sections, ×4. Same locality and collection data as GSC 126417, Pl. 15, Fig. 7.

4–6. GSC 126418, approximate topotype. Fig. 4, longitudinal section showing rhipidacanthine trabeculae, ×15; Figs. 5, 6, longitudinal sections, x4. Same locality and collection data as GSC 126417, Pl. 15, Fig. 7.

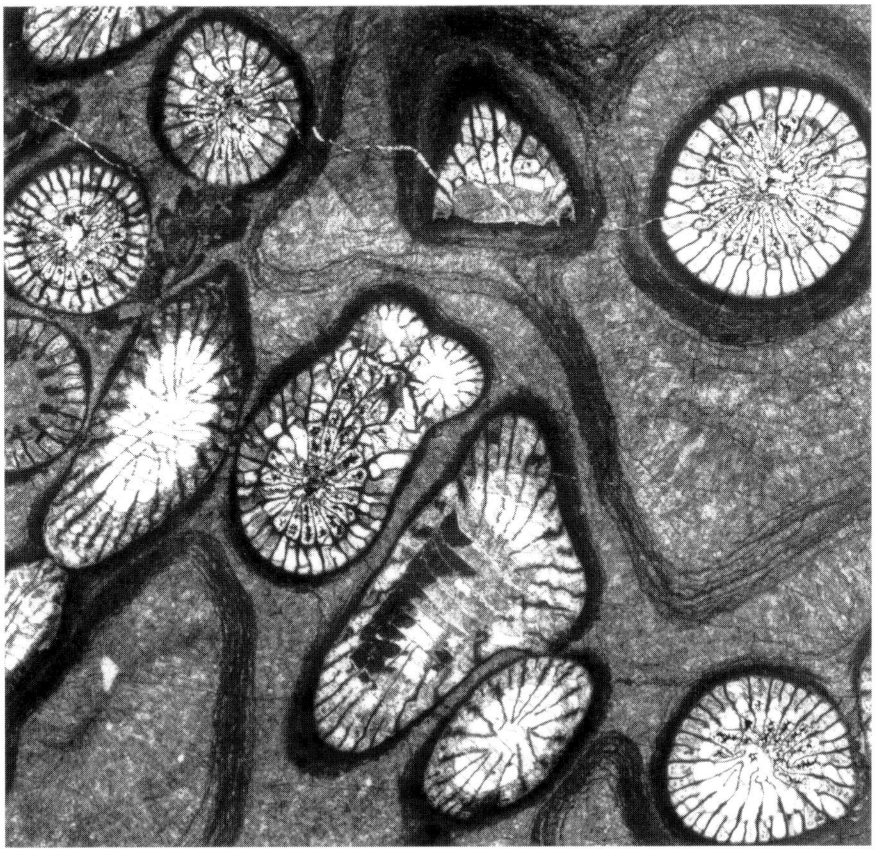

1

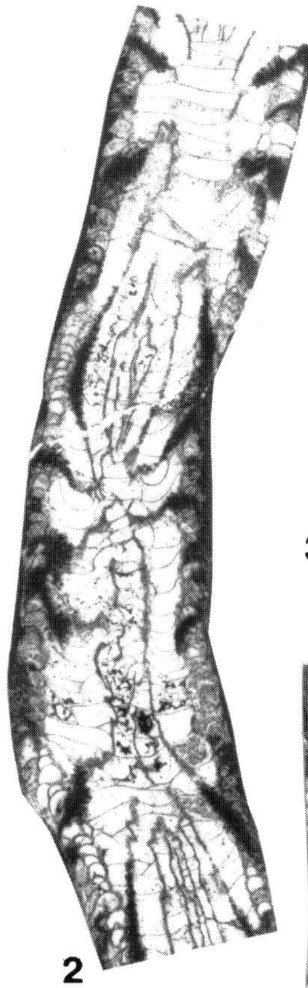

2

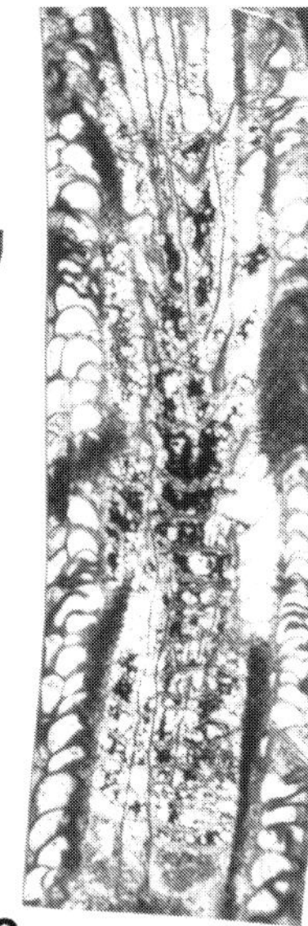

3

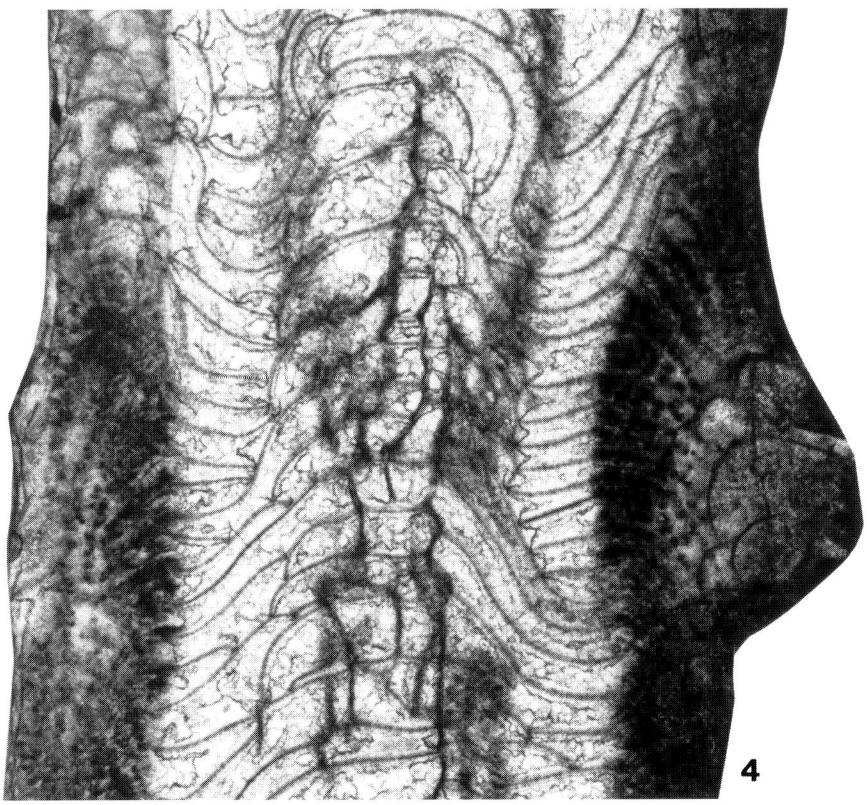

4

5

6

Plate 17

Fig. 1. *Peneckiella minor* (Roemer)

GSC 126418, approximate topotype. Longitudinal section showing rhipidacanthine trabeculae, ×15. Same locality and collection data as GSC 126417, Pl. 15, Fig. 7.

Figs. 2–5. *Peneckiella floydensis* (Belanski)

2, 5. GSC 126430. Transverse and longitudinal sections of specimen with particularly small corallites, ×4.

3, 4. GSC 126431. Longitudinal and transverse sections, ×4.

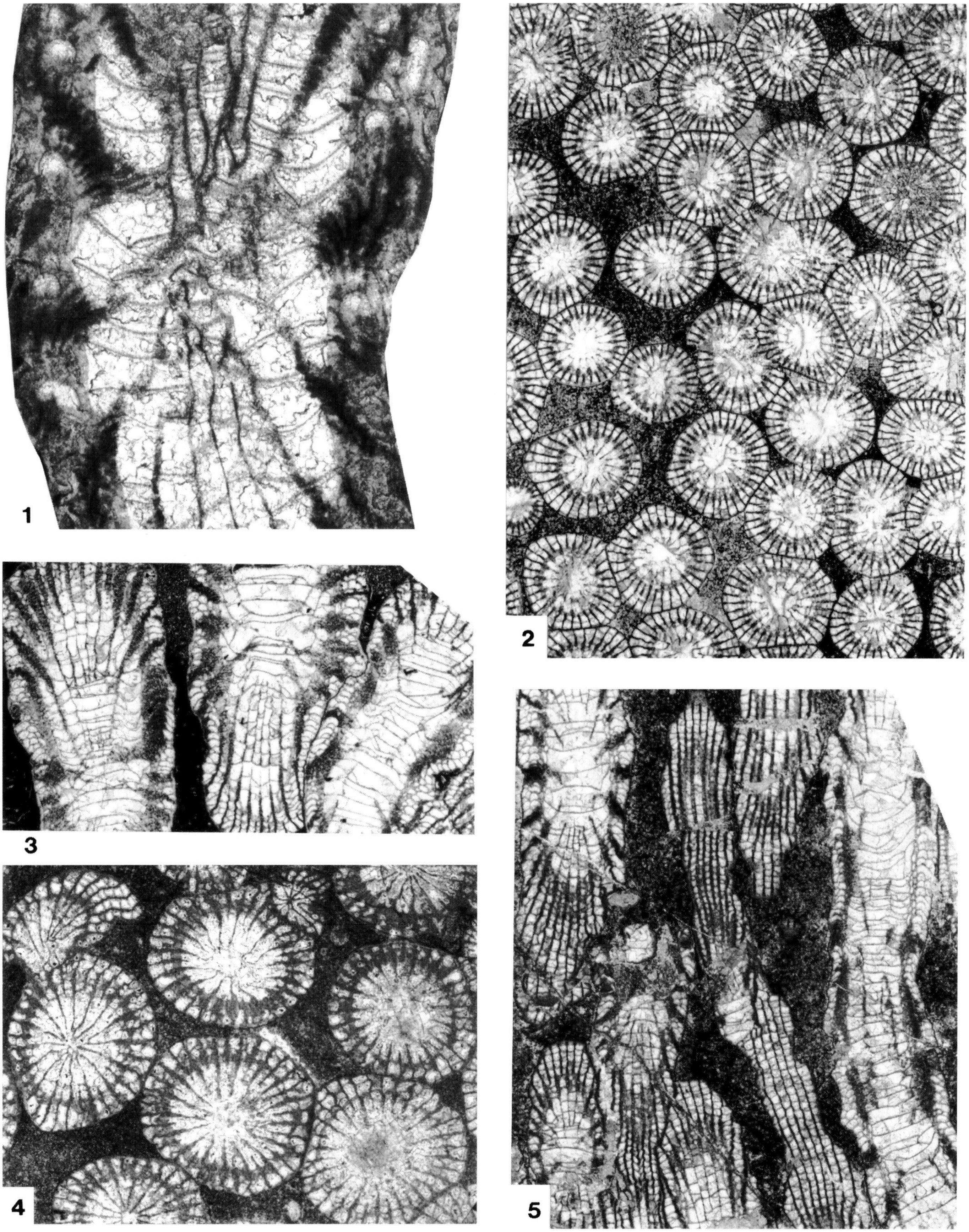
1
2
3
4
5

Plate 18

Figs. 1–6. *Peneckiella floydensis* (Belanski)

1, 2. GSC 126421. Transverse and longitudinal sections, ×4.

3, 6. GSC 126422. Transverse and longitudinal sections, ×4.

4, 5. GSC 126453. Fig. 4, longitudinal section of individual corallite, showing well-developed peneckielloid dissepiments, particularly at right, ×8; Fig. 5, transverse section, ×4. Uppermost Mason City Member, Shellrock Formation, mid Frasnian, Nora Dam, Nora Springs, Iowa (Loc. 8 of Sorauf 1998). Collected by J.E. Sorauf.

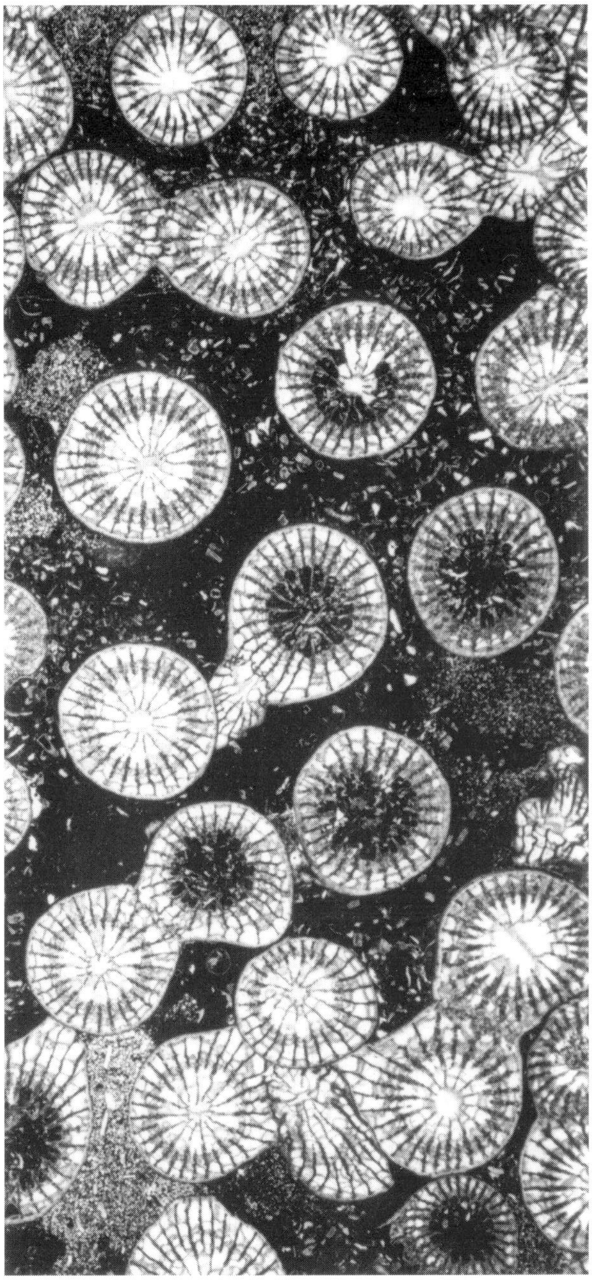

1

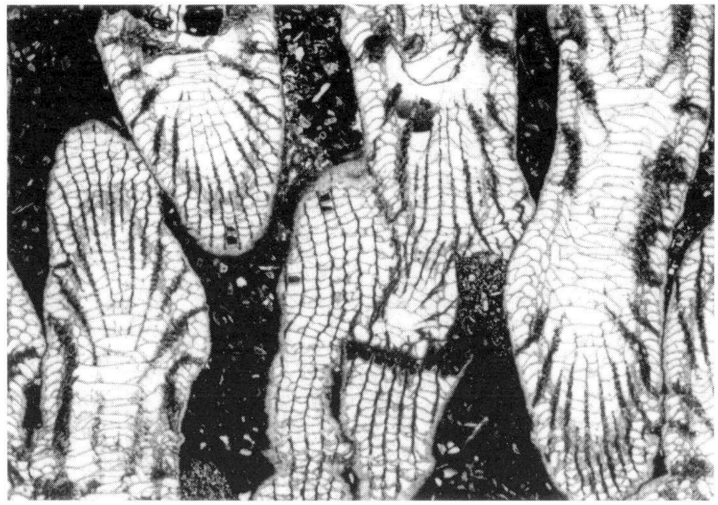

2

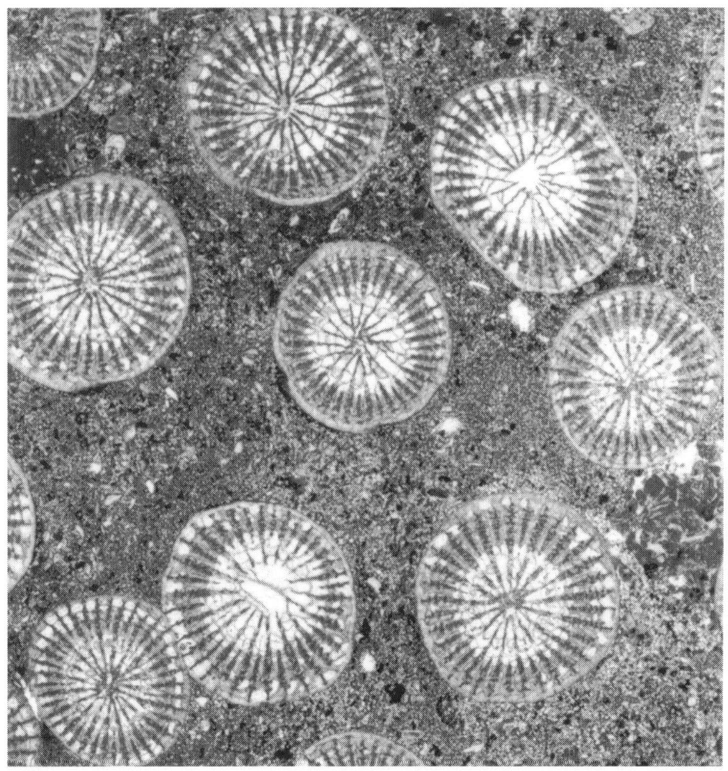

3

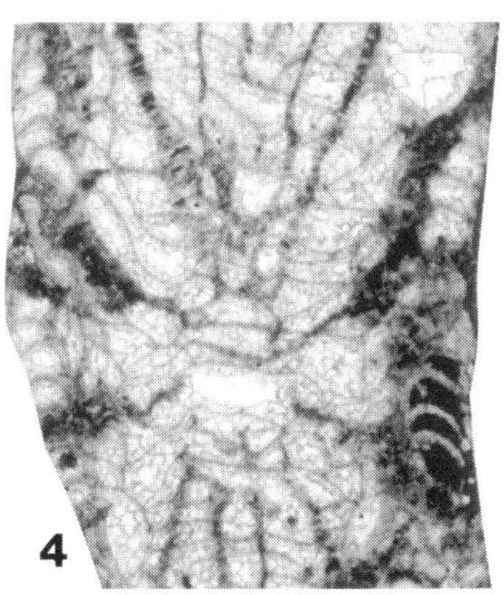

4

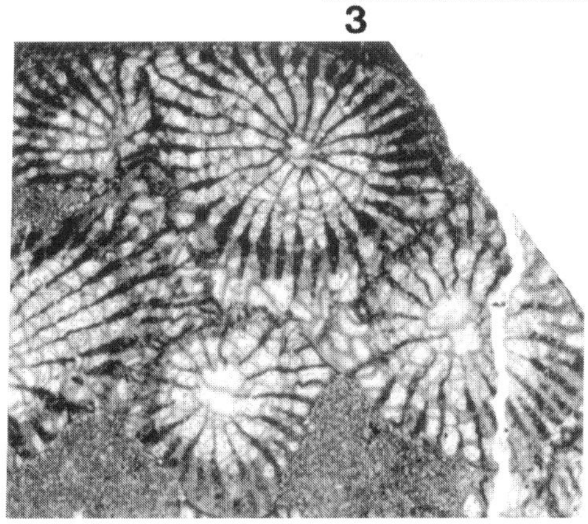

5

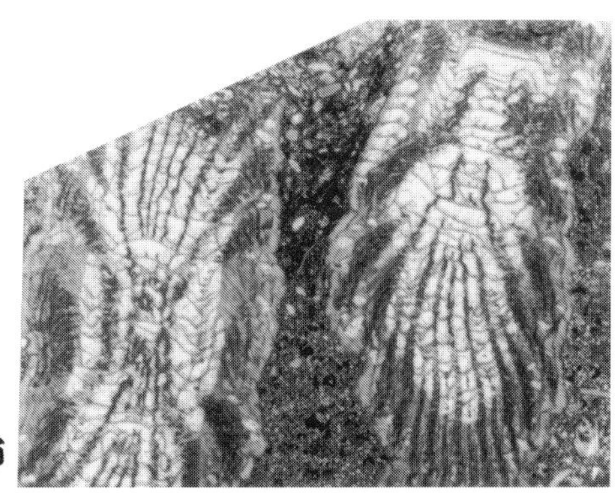

6

Plate 19

Figs. 1–7. *Peneckiella floydensis* (Belanski)

1, 3. GSC 126420. Transverse and longitudinal sections, ×4. See also Pl. 20, Fig. 4.

2, 4. GSC 6312, holotype of *Disphyllum* (*Synaptophyllum*) *densum* Smith. Transverse and longitudinal sections, ×4.

5, 6. GSC 126419. Longitudinal and transverse sections, ×4.

7. GSC 126432. Longitudinal section of corallite with unusually well-developed horseshoe dissepiments, ×4.

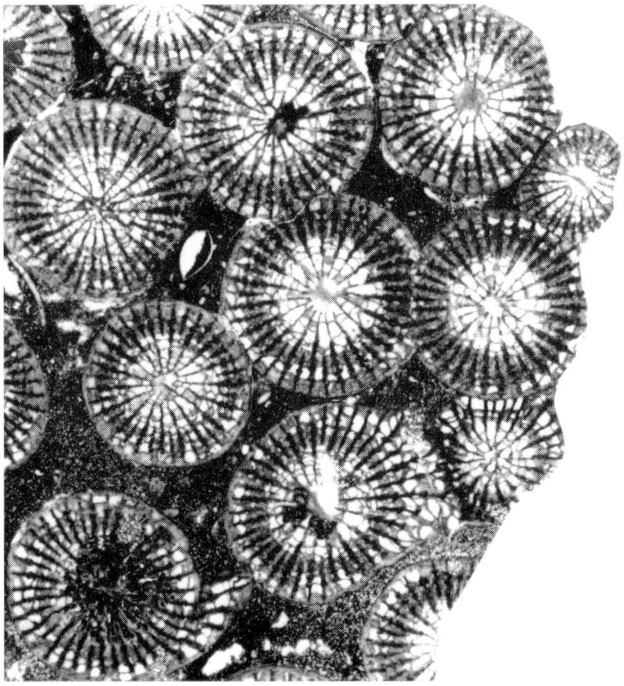

1

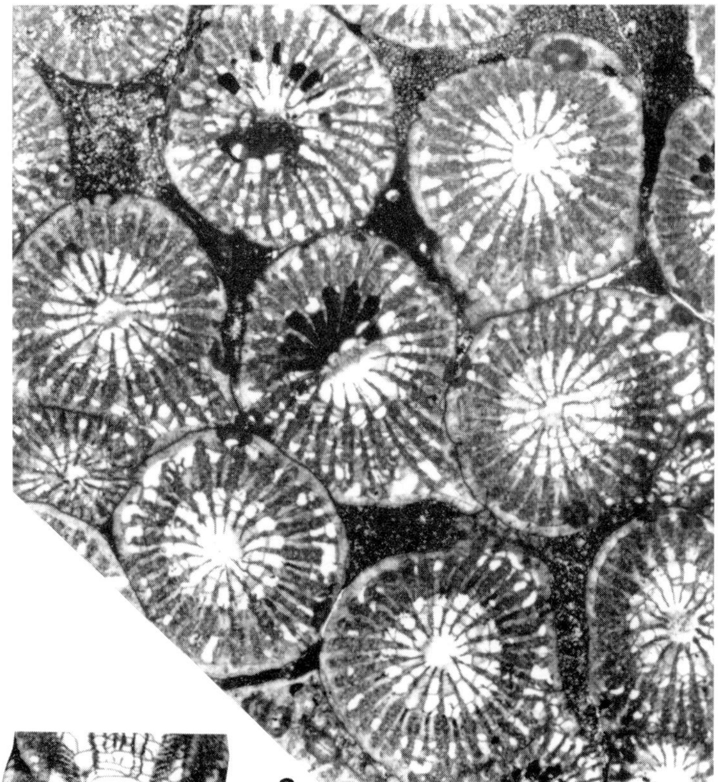

2

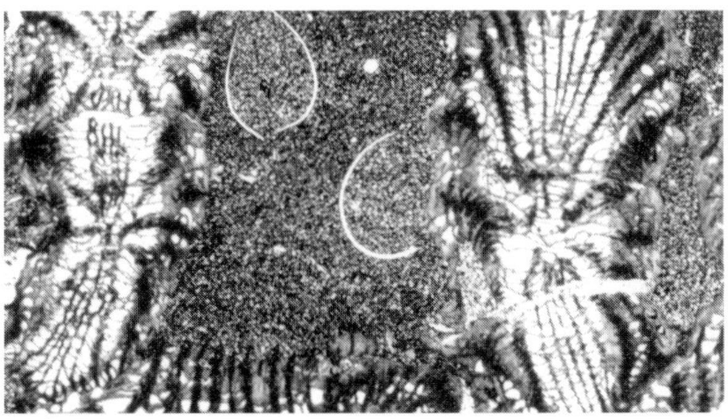

3

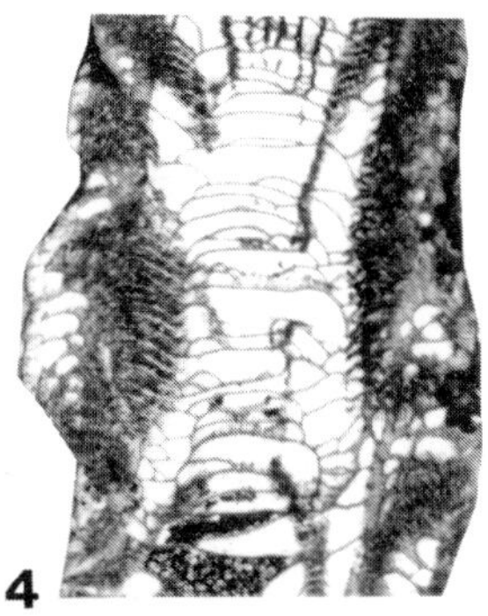

4

5

6

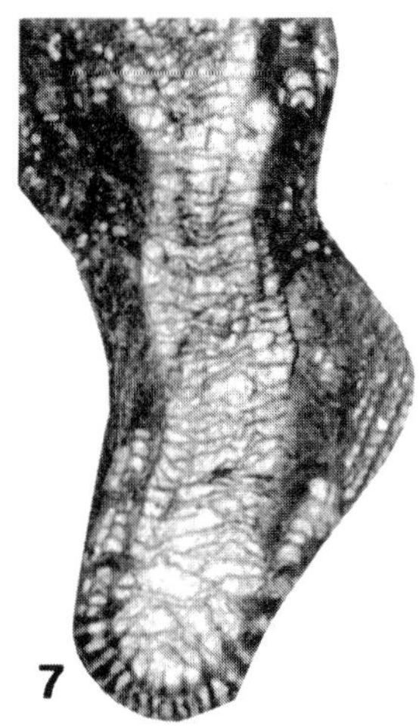

7

Plate 20

Figs. 1–4. *Peneckiella floydensis* (Belanski)

1–3. GSC 126423. Figs. 1, 2, transverse and longitudinal sections, ×4; Fig. 3, longitudinal section showing rhipidacanthine trabeculae, ×12.

4. GSC 126420. Longitudinal section showing horseshoe, peneckielloid,and globose dissepiments, and rhipidacanthine trabeculae, ×12. See also Pl. 19, Figs. 1, 3.

Figs. 5, 6. *Peneckiella metalinae* Sorauf

GSC 126439. Transverse and longitudinal sections, ×6.

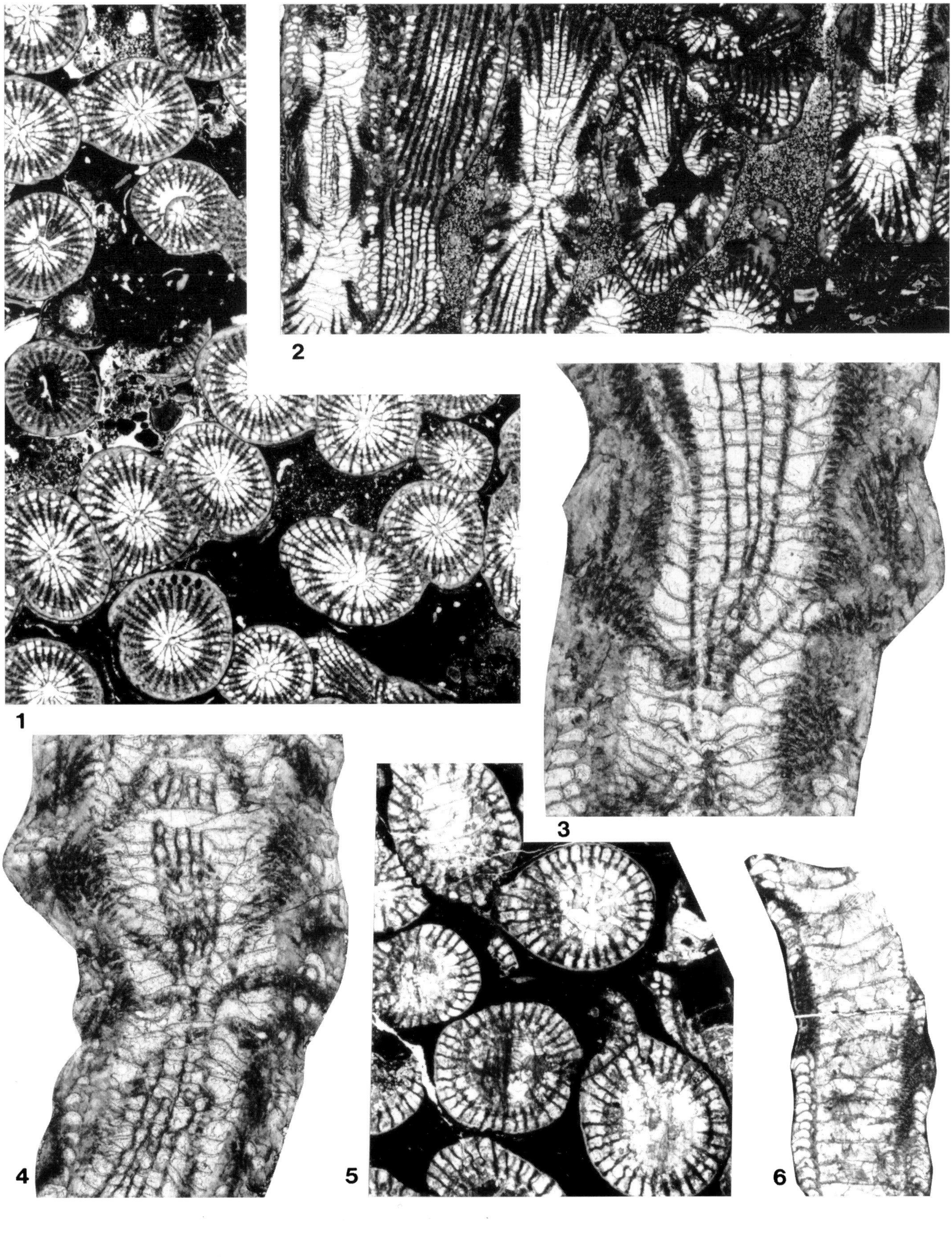
1
2
3
4
5
6

Plate 21

Figs. 1–4. *Peneckiella metalinae* Sorauf

1, 2. GSC 126433. Transverse and longitudinal sections, ×6.

3, 4. GSC 126434. Transverse and longitudinal sections, ×6.

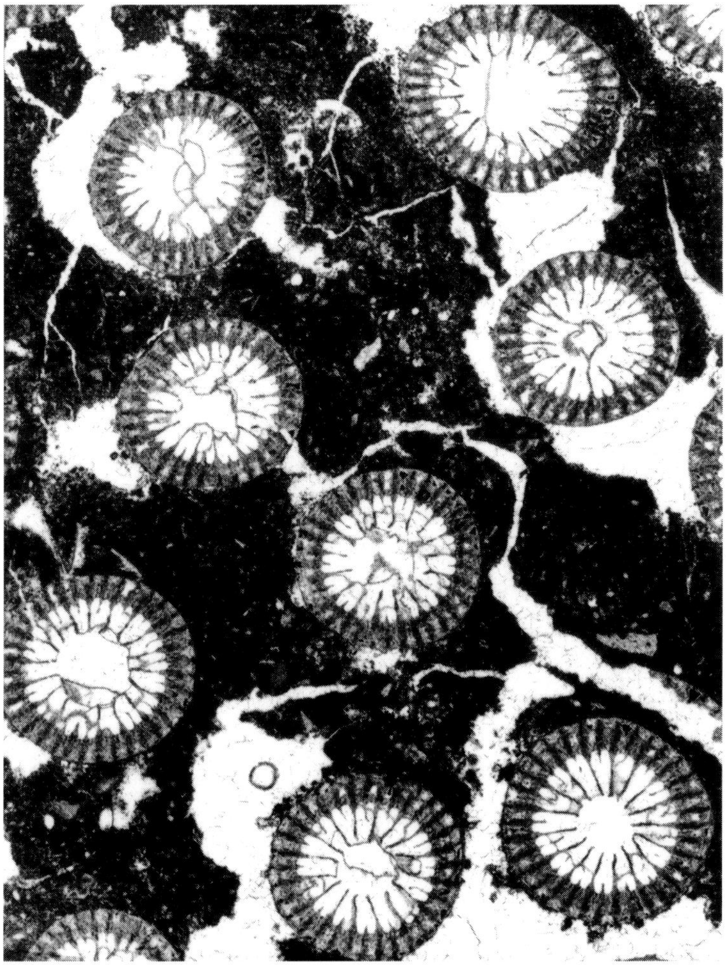

1

2

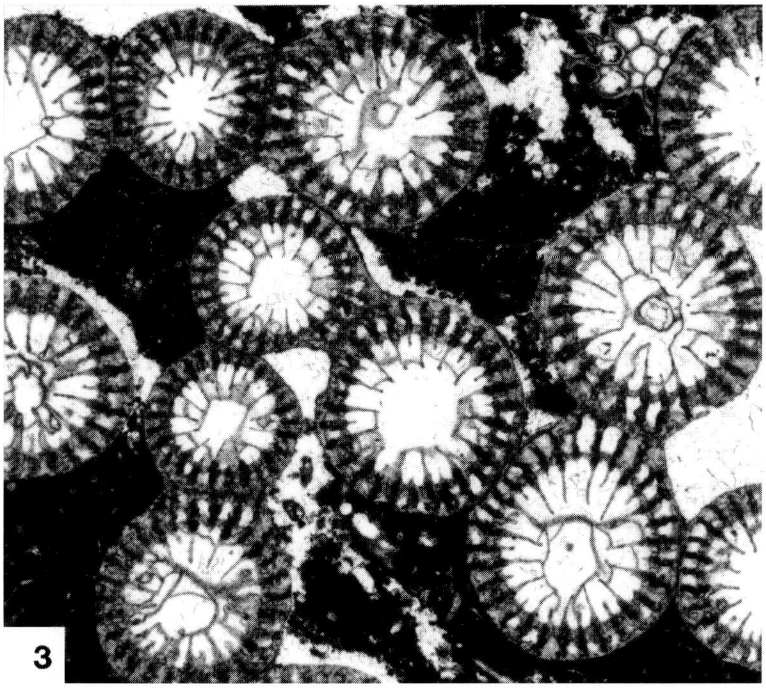

3

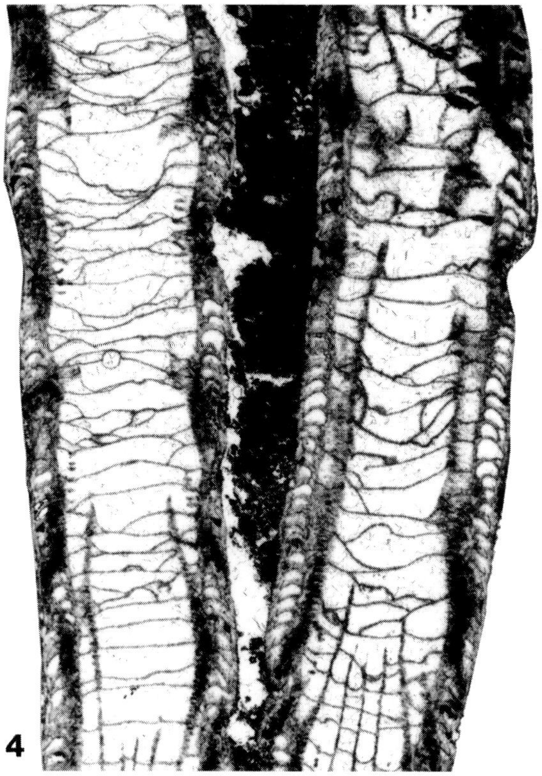

4

Plate 22

Figs. 1, 2. *Peneckiella metalinae* Sorauf

GSC 126436. Transverse and longitudinal sections, ×6. Note variation in length of major septa in Fig. 1.

Fig. 3. *Peneckiella gracilis* n. sp.

GSC 15481, *paratype*. Fragmented corallum with transverse and longitudinal sections, ×6.

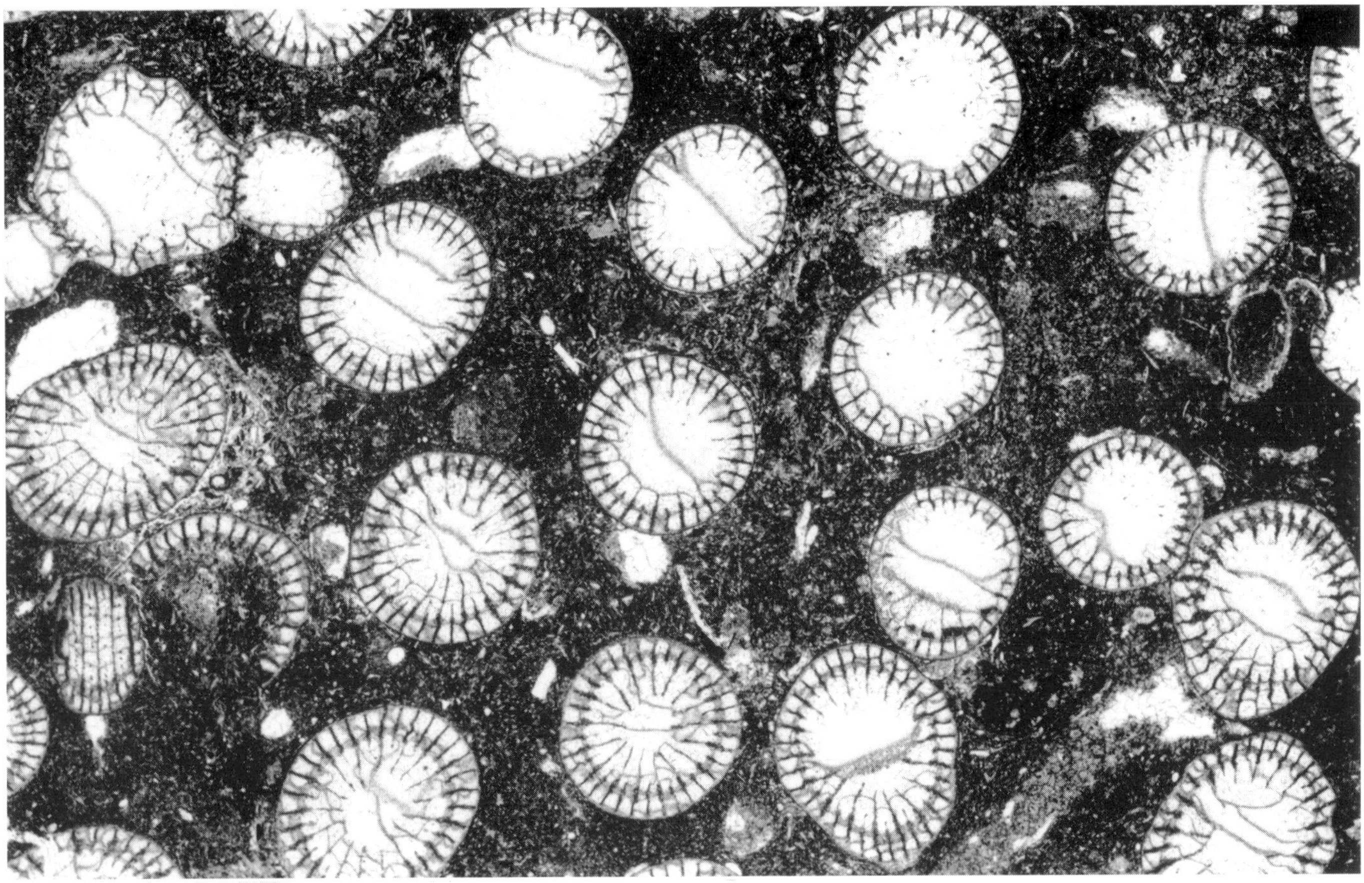

1

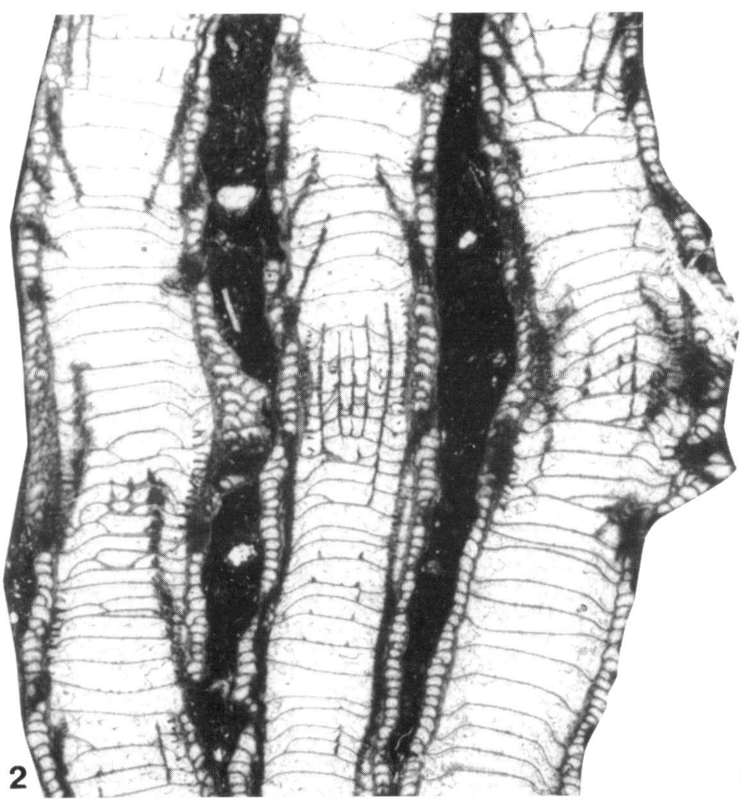

2

3

Plate 23

Figs. 1–3. *Peneckiella metalinae* Sorauf

GSC 126440. Fig. 1, longitudinal section showing rhipidacanthine trabeculae, ×15; Figs. 2, 3, transverse and longitudinal sections, ×6.

Figs. 4, 5. *Peneckiella gracilis* n. sp.

GSC 126442, *paratype*. Transverse and longitudinal sections, ×6.

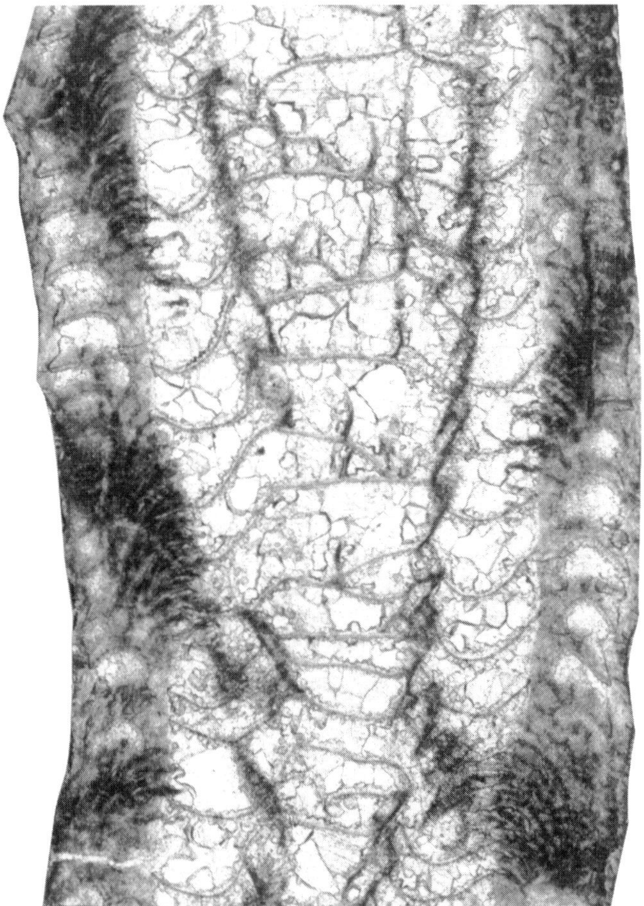

1

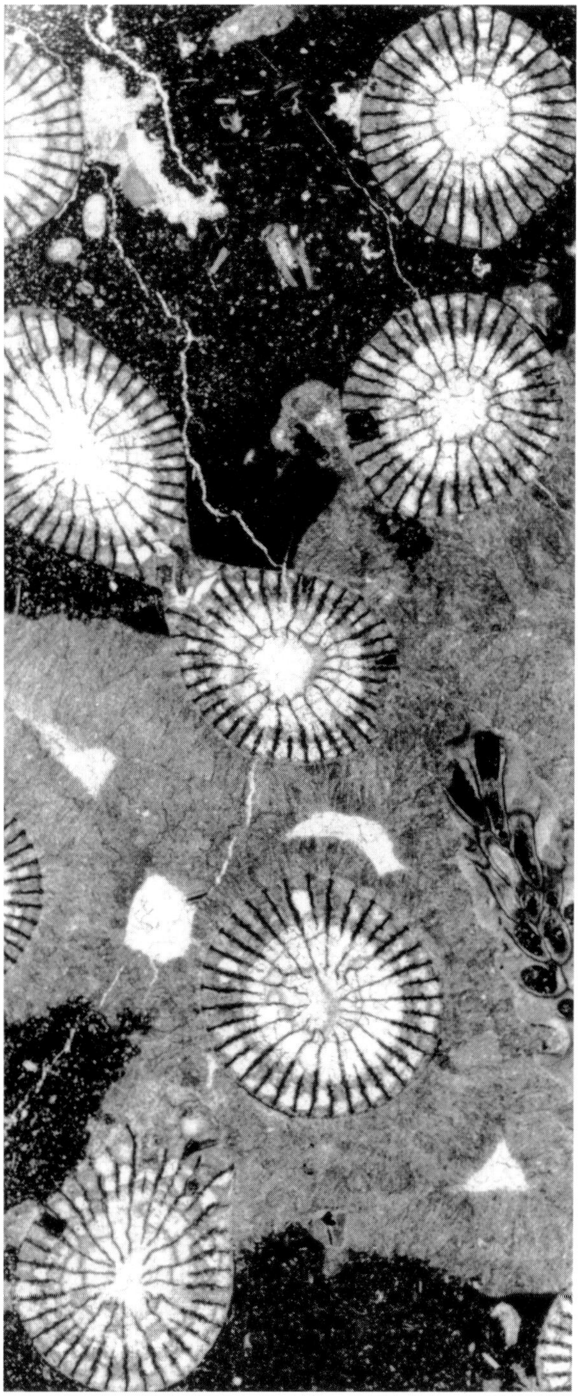

2

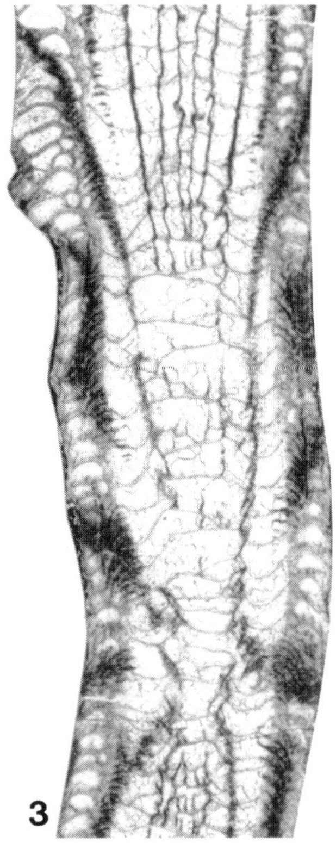

3

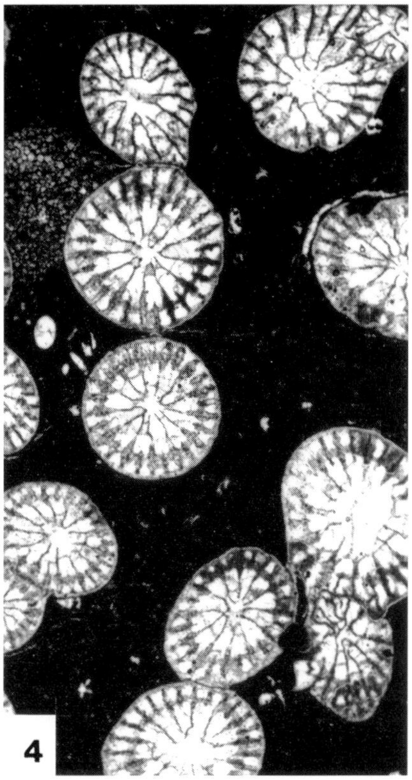

4

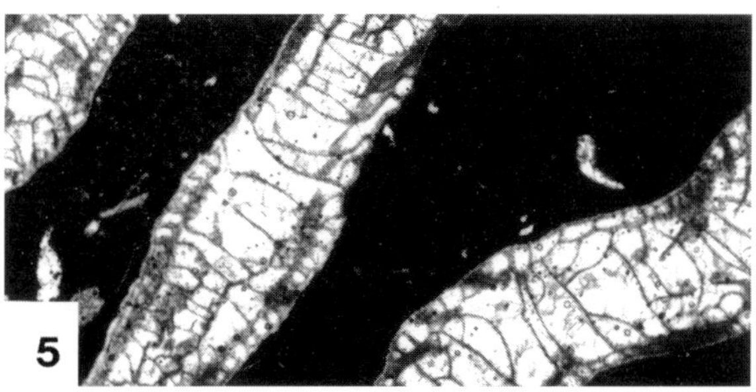

5

Plate 24

Figs. 1, 3. *Peneckiella metalinae* Sorauf

GSC 126435. Transverse and longitudinal sections of specimen with pronounced peripheral septal dilation and stereome, ×6.

Figs. 2, 4. *Peneckiella gracilis* n. sp.

GSC 126441, *holotype*. Transverse and longitudinal sections, ×6.

1

2

3

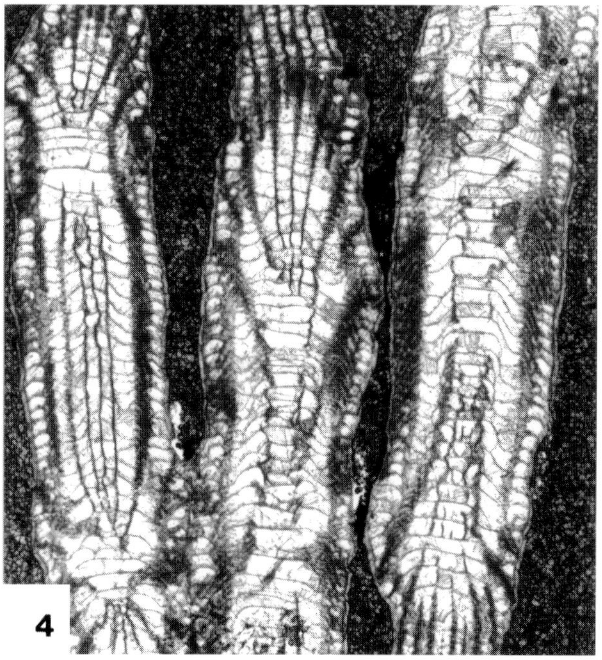

4

Plate 25

Figs. 1–8. *Peneckiella haultainensis* n. sp.

1, 2. GSC 126446, *holotype*. Longitudinal and transverse sections, ×6.

3, 4, 6. GSC 126452, *paratype*. Longitudinal and transverse sections, ×6.

5, 7. GSC 126541, *paratype*. Transverse and longitudinal sections, ×6.

8. GSC 126447, *paratype*. Longitudinal section showing horseshoe and peneckielloid dissepiments, ×15. See also Pl. 26, Fig. 6.

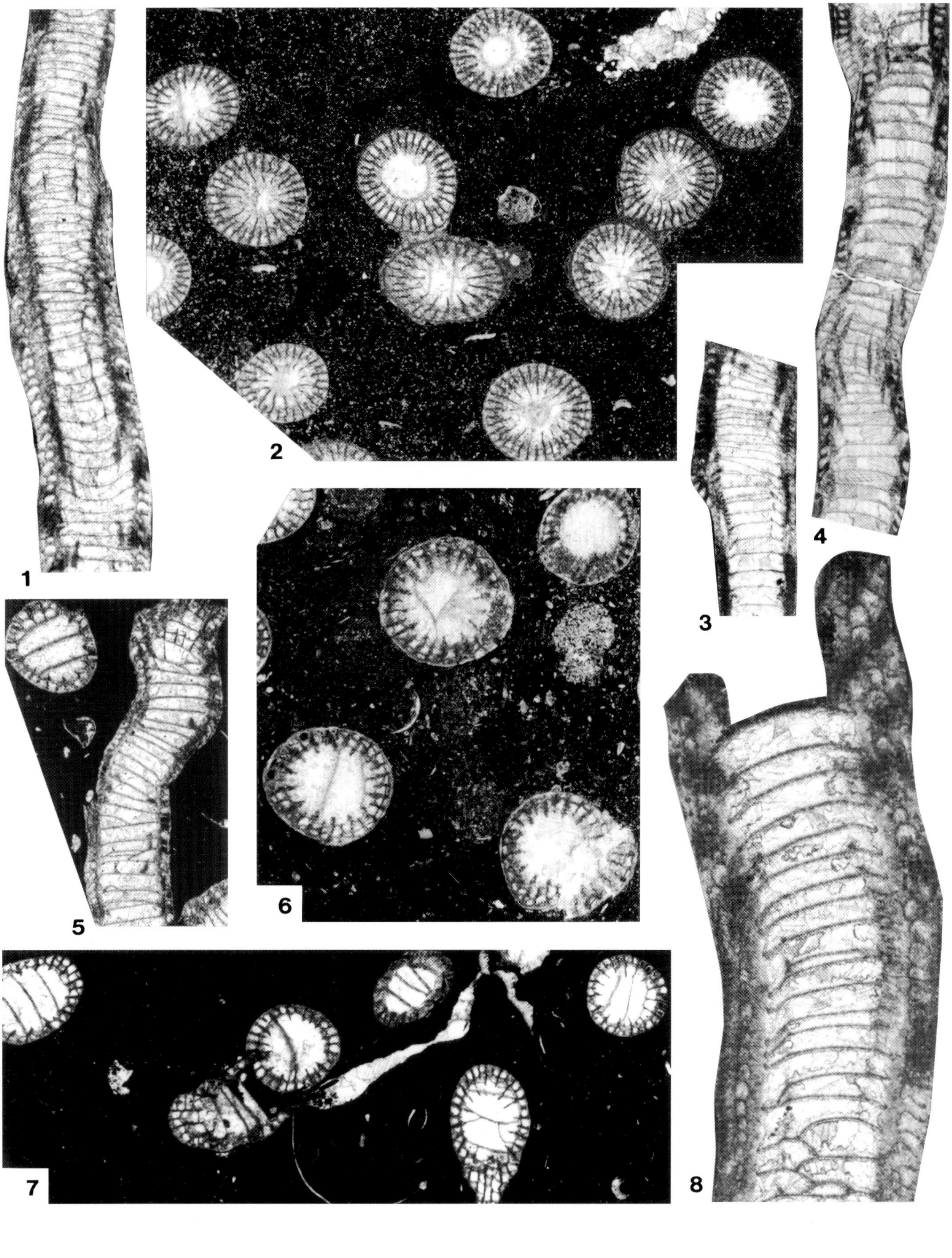

Plate 26

Figs. 1–6. *Peneckiella haultainensis* n. sp.

1, 3. GSC 126450, *paratype*. Transverse and longitudinal sections, ×6.

2, 4. GSC 126449, *paratype*. Longitudinal and transverse sections, ×6.

5. GSC 126448, *paratype*. Longitudinal section showing peneckielloid, globose and horseshoe dissepiments, and rhipidacanthine trabeculae, ×15.

8. GSC 126447, *paratype*. Transverse section, ×6. See also Pl. 25, Fig. 8.

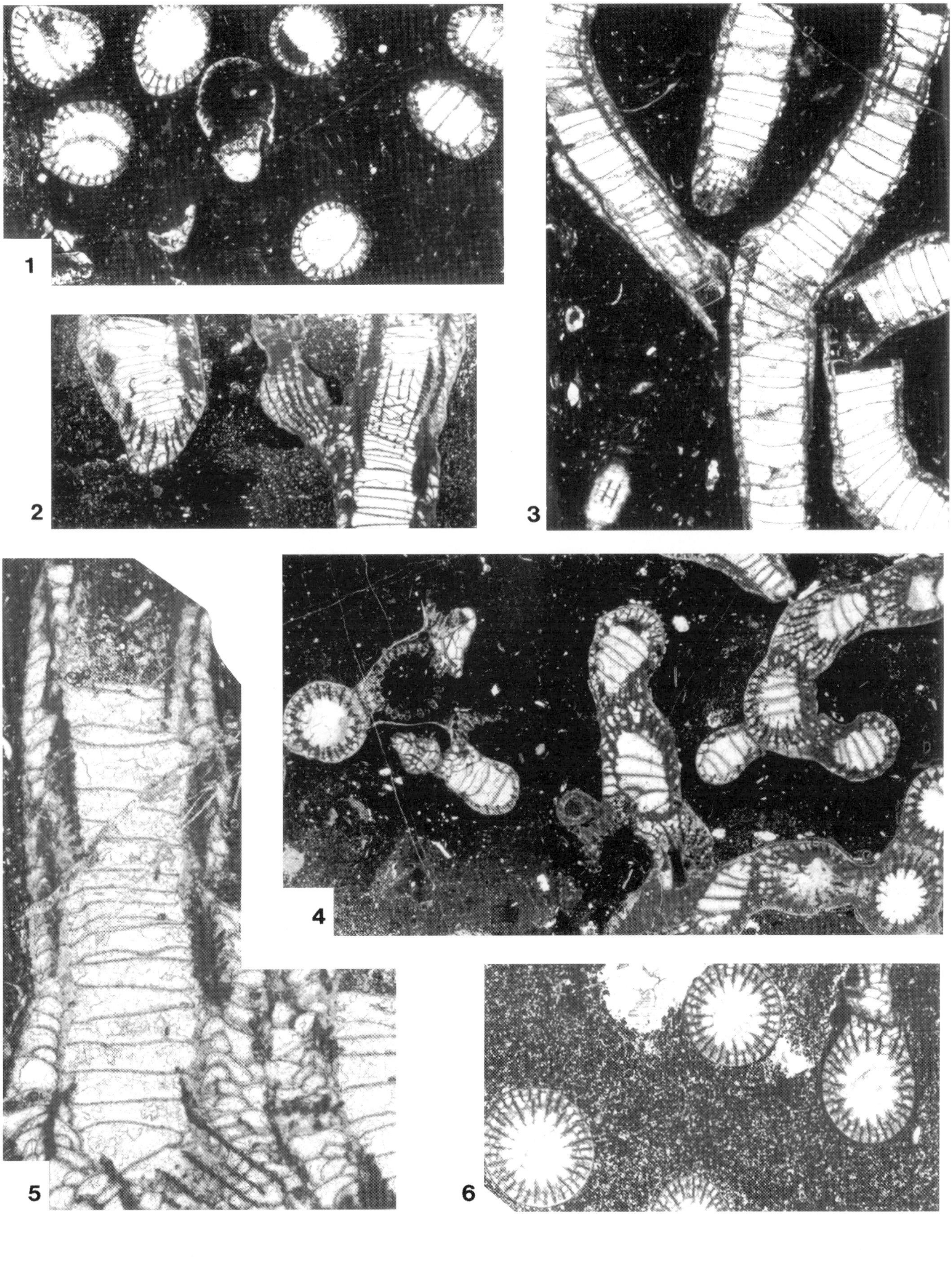
1
2
3
4
5
6